An Examination of the Methodology for Awarding Imminent Danger Pay and Hostile Fire Pay

BETH J. ASCH, JAMES V. MARRONE, MICHAEL G. MATTOCK

NATIONAL DEFENSE RESEARCH INSTITUTE

Prepared for the Office of the Secretary of Defense
Approved for public release

For more information on this publication, visit www.rand.org/t/RR3231

Library of Congress Cataloging-in-Publication Data is available for this publication.

ISBN: 978-1-9774-0353-7

Published by the RAND Corporation, Santa Monica, Calif.

© Copyright 2019 RAND Corporation

RAND® is a registered trademark.

Cover: U.S. Marine Corps photo by Cpl. Ian Ferro.

Support RAND

Make a tax-deductible charitable contribution at
www.rand.org/giving/contribute

www.rand.org

Preface

Section 612 of the National Defense Authorization Act (NDAA) for Fiscal Year 2019 directed the U.S. Department of Defense to evaluate the methodology for awarding hostile fire pay and imminent danger pay and provide a report. These pays are commonly referred to as *combat pay*. The NDAA requested information regarding the current methodology used for awarding these pays, whether it is effective in meeting the needs of service members or whether an alternative approach based on deployments would be more appropriate, and in what ways these pays could be improved to address difficulties in implementation.

In preparing its report to Congress, the Department of Defense asked the RAND Corporation to provide analytic support, and this report documents RAND's research. This report should be of interest to the defense manpower policy community and officials with responsibility for military pay policy, particularly as it relates to imminent danger pay and hostile fire pay.

This research was sponsored by the Office of the Deputy Assistant Secretary of Defense for Military Personnel Policy and conducted within the Forces and Resources Policy Center of the RAND National Defense Research Institute, a federally funded research and development center sponsored by the Office of the Secretary of Defense, the Joint Staff, the Unified Combatant Commands, the Navy, the Marine Corps, the defense agencies, and the defense Intelligence Community.

For more information on the RAND Forces and Resources Policy Center, see www.rand.org/nsrd/ndri/centers/frp or contact the director (contact information is provided on the webpage).

Contents

Figures and Tables

Summary

Hostile fire pay (HFP) and imminent danger pay (IDP) are two of the pays that the U.S. Department of Defense (DoD) uses to recognize the risks faced by service members in the line of duty. Both pays are $225 per month, and IDP is prorated to the number of days served in a qualifying area at a rate of $7.50 per day. Members cannot receive HFP and IDP simultaneously. HFP is event-based and is paid when members are exposed to hostile fire. IDP is location-based and is paid when a member is performing duties in a foreign area where he or she may be subject to the threat of physical harm or imminent danger on the basis of civil insurrection, civil war, terrorism, or wartime conditions. These pays are commonly referred to as *combat pay*. Service members exposed to danger may receive other pays, such as parachute pay, and overall military pay levels partially reflect the arduous duties that military service entails. The focus of the research summarized in this report is on HFP and IDP, which we compare with other risk-based pays used by the U.S. military, federal and international agencies, and foreign militaries.

Section 612 of the National Defense Authorization Act (NDAA) for Fiscal Year (FY) 2019 directed DoD to evaluate the methodology for awarding HFP and IDP and provide a report. The congressional language raised concerns that the methodology for awarding HFP and IDP may no longer serve the needs of members in the current operational environment. The FY 2019 NDAA requested information regarding the current methodology used for awarding these pays, whether it is effective in meeting the needs of service members or whether an alternative approach based on deployments would be more appropriate, and in what ways these pays could be improved to address difficulties in implementation. In particular, the congressional language evinced concern about the current geographic model for awarding pay to members, indicating that the concern was primarily with IDP rather than HFP. The Office of Compensation within the Office of the Under Secretary of Defense for Personnel and Readiness engaged the RAND Corporation to help gather the information and perform the analysis necessary to assist DoD in its response to Congress. This report summarizes our findings.

To address these questions, we reviewed policy documents; past studies; the history of risk-based pay in the United States; and the use of such pays in other organizations, including the federal civil service and foreign militaries. In addition, we interviewed subject-matter experts in the services, the Office of the Secretary of Defense, the Joint Staff, the combatant commands (COCOMs), and other interested parties. We used these inputs to address four broad questions that we formulated to cover the issues raised by the FY 2019 NDAA language:

1. What is the current methodology for making an HFP or IDP designation or recommendation?
2. Is the current IDP process effective? Does it meet the needs of service members, including special operations forces?

3. Is geographic eligibility the best method for awarding IDP, or would a different approach, particularly one based on deployment or operations, be more effective and efficient?
4. What are the difficulties in implementing the current system?

Findings

The Current Methodology Is Effective but Not Considered Efficient

The current methodology for designating HFP and IDP is based on DoD Instruction (DoDI) 1340.09, *Hazard Pay (HzP) Program* (DoD, 2018). Almost all of the interviewees indicated that the HFP and IDP process is administratively tractable and provides pay to those who are serving in a threatening environment. Furthermore, interviewees were not aware of people falling through the cracks and not eventually receiving the pay they were due. And none of the studies we reviewed indicated that service members serving in a threatening environment did not receive HFP or IDP.

In some cases, modern warfare and a more dynamic threat environment may have led to some personnel, such as special operations forces, missing out on IDP, given that their missions are often classified and they are rapidly deployed. However, subject-matter experts noted that other pay mechanisms are in place to recognize these service members. First, as an event-based pay, HFP is available if members are subject to hostile fire. Second, in the case of special operations forces, personnel qualify for other special and incentive pays. Finally, the dangerous and arduous aspects of military service are essentially priced into basic pay. Research shows that regular military compensation is above the average compensation, and indeed above the 70th percentile of civilians with similar characteristics (Hosek et al., 2018).

Our interviewees indicated that the IDP certification process is typically quite lengthy and thus does not respond to changes in the threat environment in a given location in a timely way.[1] The long process is driven by the number of offices that have a role in or responsibility for IDP designation. DoDI 1340.09 requires periodic review of the countries designated as eligible for IDP but does not stipulate the frequency of such reviews. Such reviews have occurred periodically, but many of the interviewees raised concerns about how long some countries have been on the IDP list; the "bad optics" this can create, especially if the threat to civilians appears to be low; and the cost to the services of IDP for locations where the risk is low.

Many interviewees deemed the $225-per-month rate for HFP and IDP to be too low. We could find no evidence of a connection between this pay and recruiting and retention outcomes. But, insofar as HFP and IDP recognize dangerous duty, the value has eroded over time because the dollar amount has remained unchanged since 2003. Furthermore, HFP and IDP are less than the Family Separation Allowance amount of $250 per month. Some interviewees requested better communication within DoD about the progress of IDP packages to give visibility to commanders about when soldiers, sailors, airmen, and marines can expect to receive retroactive IDP payments. Some also requested better access to up-to-date information on the current list of IDP-designated locations. Finally, the definition of threat for designating IDP focuses on security risk, acts of violence toward service members, wartime conditions, and the like, but not health risks.

---

[1] Although the median time for the certification process between 2008 and 2018 was only six months (Office of the Under Secretary of Defense for Personnel and Readiness, 2018), one interviewee mentioned two packages that had been in process for years without being resolved.

The Methodology Could be Improved by Allowing Imminent Danger Pay Rates to Vary with the Severity of Threat but Not by Basing Them on Deployments

Several interviewees said that the severity of threat can vary widely both within an IDP-designated location and across locations. Furthermore, the Eleventh Quadrennial Review of Military Compensation (QRMC) analyzed metrics of danger and found that the correlation between combat compensation and degree of danger had eroded (Office of the Under Secretary of Defense for Personnel and Readiness, 2012). The Eleventh QRMC and several interviewees recommended multiple tiers of IDP rates that would be correlated with the severity of threat. IDP would still recognize personnel exposed to lower levels of threat, but the pay rate would be lower than the rate for those exposed to higher levels of threat.

In our review of how other organizations, including foreign militaries, award risk-related pay, we found that nearly all of the organizations awarded pay on the basis of location only, although the pay rate may vary by location. The exception was the Canadian military, which bases pay on mission risk and location. Missions are assigned a risk level from I to IV, and the amount of danger pay varies with risk level. Thus, the Canadian military's experience shows that there is some precedent for a danger-related pay that varies with severity of risk.

Interviewees uniformly rejected the idea of basing IDP on deployments regardless of location, as well as on deployments associated with hazardous locations. Deployments do not necessarily align with risk, so these approaches would not be fair to service members. Deployed service members who are facing no more risk than when they are in the United States would receive more pay, while those who are in danger but are not considered deployed would not receive the pay. Furthermore, such approaches were deemed infeasible, and maybe even impossible, to implement. Tracking individual locations and missions was considered extremely burdensome from an administrative standpoint. Finally, interviewees questioned the need for a deployment-related pay, because HFP compensates for exposure to an actual event and can be originated by commanders in the field. Consequently, these interviewees argued that DoD already has a mechanism built into the HFP and IDP process to recognize exposure to danger that is not geographically based.

Recommendations

From these findings, we make the following recommendations. Some of these recommendations would require amending the DoDI, while others would require congressional action.

Create tiered rates of IDP based on severity of threat. Setting IDP to reflect different levels of exposure to danger would address inequities among members who currently receive the same pay but face different exposure. It would also respond to the Eleventh QRMC's finding that the connection between danger and IDP has eroded. One potential concern about such a system is that it would require defining *severity of threat*, although interviewees told us that DoD already makes assessments of severity of threat. Another concern is that multiple tiers within an area of responsibility would exacerbate the "haves and have nots" challenge, which in turn might exacerbate the certification timeliness problem. Tracking individual locations within an area of responsibility may also prove burdensome. Another potential concern is that IDP costs could increase. One way to address cost is to set the lowest level of IDP for the least threat below the current $225 per month. Creating tiered rates of IDP based on severity of threat would require congressional action.

Increase the current $225 rate for HFP and IDP. IDP should be increased to restore its real value since 2003 and to exceed the $250-per-month Family Separation Allowance, at least for some members. Such a policy could increase costs. If increasing HFP and IDP is not feasible from a budgetary perspective, an alternative could be to restructure the pays in a cost-neutral way but still base IDP on severity of threat. For example, as part of a tiered IDP system, the lowest-risk IDP rate could fall below $225, while the higher-risk IDP rates could exceed $250 per month. Increasing the IDP rate above $250 would require congressional action because Congress has authorized the pay only up to $250. Increasing the maximum authorized HFP beyond $450 would similarly require congressional action.

Identify whether it is possible to reduce the length of time for IDP certification. IDP certification requires input from a large number of stakeholders, and although any given stakeholder may have limited time to provide input, the overall process is lengthy. DoD should map out the process and assess how long each stage requires for concurrence to identify whether there are ways to streamline the process—for example, by doing some steps concurrently or even eliminating some steps altogether. Identifying whether it is possible to reduce the length of time for IDP certification would require DoD action.

Institutionalize regular periodic reviews of IDP designations. The last worldwide review was in 2014, and a new one is being completed; thus, arguably, reviews could occur regularly every five years. An alternative approach is to include a sunset provision that would decertify a location after five years, unless an HFP event or other information indicates that the location continues to be a threat to service members. Although the periodic reviews do occur and the combatant commanders have the flexibility to initiate designation and removal of IDP locations, the commanders' incentives to remove designations may not always align with that of the services. A periodic review would also provide an ongoing basis for justifying the current list of IDP designations. Implementing this recommendation would save money if IDP was no longer paid after the threat at a location has diminished. Institutionalizing a regular periodic review of IDP designations would require DoD action.

Amend DoDI 1340.09 to require the geographic COCOMs to seek input and concurrence from the special operations commander within each COCOM's area of responsibility on packages relevant to special operations forces. This input would occur before the IDP designation request was submitted to the Joint Staff. Doing so would help ensure that information that might currently be missed would be incorporated in the IDP designation request as a matter of course at the point of origination within the geographic COCOMs. Amending DoDI 1340.09 would require DoD action.

Review the criteria in DoDI 1340.09 to assess whether additional risks to service members should be considered as a criterion for designating IDP. For example, biological risks to service members who might be exposed during biological warfare might be added to the list of designation criteria. Insofar as exposure to biological hazards is an aspect of duty rather than a result of hostilities, DoD should consider allowing such members to qualify for hazardous duty pay. Reviewing the criteria would require DoD action.

Create a capability that would allow IDP administrators across DoD to access up-to-date information. Such a capability could be a website accessible to those involved in the determination process or administration of IDP, or it could be a monthly or quarterly update sent by email to the relevant individuals. This communication would ensure that all involved in the determination process or administration of IDP have a central, authoritative source for IDP information, particularly an up-to-date list of designated countries. Creating this capability would require DoD action.

Acknowledgments

We are indebted to the military personnel management experts who participated in the interviews we conducted. Their input was critical to our study. We are grateful to Tim Fowlkes in the Office of Compensation within the Office of the Under Secretary of Defense for Personnel and Readiness. Fowlkes served as project monitor, provided background material, and arranged interviews for the project. We are also grateful to Jerilyn Busch, director of that office, who provided input and guidance for our analysis. At RAND, we wish to thank Tina Panis and Kurt Klein for their help with our project. We are also grateful for the input and comments from the report's two reviewers, Joshua Klimas at RAND and Patrick Mackin at SAG Corporation.

Abbreviations

AFRICOM	U.S. Africa Command
ASD(M&RA)	Assistant Secretary of Defense for Manpower and Reserve Affairs
CENTCOM	U.S. Central Command
CCDR	combatant commander
COCOM	combatant command
CTS	Contingency Tracking System
CZTE	combat zone tax exclusion
DEA	Drug Enforcement Agency
DMDC	Defense Manpower Data Center
DoD	U.S. Department of Defense
DoDI	DoD Instruction
DSSR	Department of State Standardized Regulation
EUCOM	U.S. European Command
FBI	Federal Bureau of Investigation
FY	fiscal year
GAO	U.S. Government Accountability Office
HDIP	hazardous duty incentive pay
HDP-L	hardship duty pay – location
HFP	hostile fire pay
IDP	imminent danger pay
INDOPACOM	U.S. Indo-Pacific Command
NDAA	National Defense Authorization Act
OUSD(P&R)	Office of the Under Secretary of Defense for Personnel and Readiness

QHDA   qualified hazardous duty area

QRMC   Quadrennial Review of Military Compensation

SOCOM   U.S. Special Operations Command

SOUTHCOM   U.S. Southern Command

SME   subject-matter expert

UN   United Nations

CHAPTER ONE

Introduction

Military personnel engage in dangerous duties in threatening environments. One of the key ways that the U.S. Department of Defense (DoD) recognizes these duties is by offering special pays. This report focuses on two of the special pays designed to recognize hazardous duty: hostile fire pay (HFP) and imminent danger pay (IDP). These pays are commonly referred to as *combat pay*. In brief, IDP is a monthly payment of up to $225 for members serving in a designated combat zone or in an area designated as an imminent danger area; IDP is prorated at $7.50 per day. HFP also pays $225 per month and is authorized by the same statute as IDP, but it is not prorated.[1] HFP differs from IDP in that HFP is an event-based ("the bullets are flying") pay, while IDP is a location-based pay in which payment is based on an assessment that danger of hostilities to service members exists in a particular location. Chapter Two provides more details on these pays. Later in the report, we provide a detailed comparison of the eligibility criteria and benefit levels for HFP and IDP with those of other risk-based pays used by U.S. military personnel and federal civilians, as well as by United Nations (UN) staff and foreign militaries (see Table 5.1).

Section 612 of the National Defense Authorization Act (NDAA) for Fiscal Year (FY) 2019 directed DoD to evaluate the methodology for awarding HFP and IDP and provide a report (Pub. L. 115-232). Specifically, Congress asked for analysis of how geographic regions are selected for eligibility for IDP, the criteria that are used to define those regions, and any difficulties in implementing the current HFP and IDP system. Congress also asked whether the current geographic model is the most appropriate way to award IDP and whether it would be appropriate to tie IDP to specific authorizations for deployments (including deployments of special operations forces) in addition to geographic criteria. The Office of Compensation within the Office of the Under Secretary of Defense for Personnel and Readiness (OUSD(P&R)) engaged the RAND Corporation to help gather the information and perform the analysis necessary to assist DoD in its response to Congress. This report summarizes our findings.

The reason raised in the FY 2019 NDAA for reviewing the IDP process was that the current geographic model for awarding IDP might not accurately reflect the realities of modern warfare and might not be responsive enough to the needs of service members. The risk environment is more dynamic than in the past in terms of locations and the nature of the threat to U.S. forces. An indication of this change is the growth in the use of special operations forces in recent years (Congressional Research Service, 2019). With these changes, the question arises of whether the current methodology for awarding IDP is responsive to the needs of service

---

[1] Statutory authority exists for prorating HFP for members who spend time in hostile fire areas, but this authority does not seem to be used in practice.

members; in particular, are people who are subjected to risk failing to qualify for risk-related pay or seeing a substantial delay in receiving risk-related pay? A related question raised by the NDAA asks whether tying these payments to deployment would address any gaps in the current methodology.

Our approach to assisting DoD with its response to Congress involved reviewing past studies of combat pay, as well as the methodology for designating hazardous duty pay in other organizations, including foreign militaries. We also used available data from the Defense Manpower Data Center (DMDC) to tabulate the relationship between hazardous duty pay and (1) deployments and, to the extent possible, (2) location. And we reviewed the history of combat and risk-based pay. A major element of our study was structured discussions with subject-matter experts (SMEs).[2] We also interviewed SMEs from the services, the Office of the Secretary of Defense, the Joint Staff, and the combatant commands (COCOMs). The approach focused on addressing four broad questions that cover the issues raised by the FY 2019 NDAA language:

1. What is the current methodology for making an HFP or IDP designation or recommendation?
2. Is the current IDP process effective? Does it meet the needs of service members, including special operations forces?
3. Is geographic eligibility the best method for awarding IDP, or would a different approach, particularly one based on deployment or operations, be more effective and efficient?
4. What are the difficulties in implementing the current system?

Addressing the issue of effectiveness raises the additional question of how effectiveness is judged. Our main criterion for judging effectiveness is whether HFP and IDP are meeting their stated purposes. As discussed in the report of the Tenth Quadrennial Review of Military Compensation (QRMC) (OUSD(P&R), 2008), as well as in Hosek, Mattock, and Asch, 2019, the overarching purpose of these combat pays is to act as insurance for service members against unpredictable and dangerous events. In particular, these pays should provide recognition to service members for situations in which they are unpredictably or involuntarily in harm's way. Consistent with the concept that these pays act as insurance, the level of pays should vary with the degree of risk. Furthermore, the pays should be efficiently implemented in a timely and transparent way.

Before presenting the results of our analysis, it is useful to put HFP and IDP in context with the rest of the military compensation system, including other special pays and benefits. First, DoD provides two other risk-based pay and benefits with zonal eligibility—namely, hardship duty pay – location (HDP-L) and the combat zone tax exclusion (CZTE). As three distinct types of benefits,[3] HFP and IDP, CZTE, and HDP-L have eligibility criteria that partially but not completely overlap. Consequently, a given service member may earn anywhere from zero to all three benefits. Figure 1.1 illustrates the possible types of overlap, providing examples of locations in which a service member may be eligible for different combinations of

---

[2]   RAND's institutional review board determined that our study was not research involving human subjects.

[3]   In this comparison, we group HFP and IDP together because, as explained in Chapter Two, IDP can be understood as a subset of HFP. It is common to refer to (1) HFP and IDP and (2) HDP-L as special *pays*, but because CZTE is not a pay, we use the term *benefit* to characterize all three when referring to them together.

Figure 1.1
Eligibility Overlap Between HFP and IDP, HDP-L, and CZTE, as of June 2019

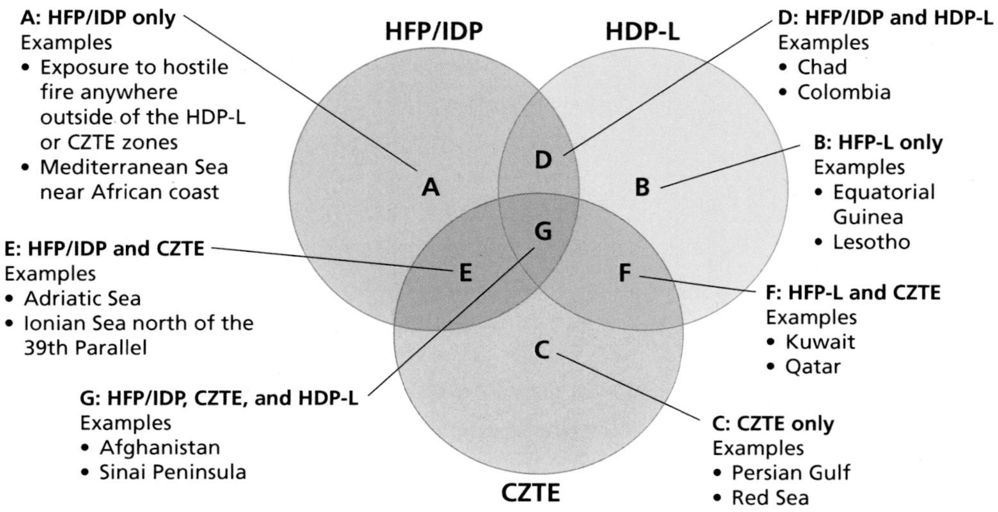

A: HFP/IDP only
Examples
- Exposure to hostile fire anywhere outside of the HDP-L or CZTE zones
- Mediterranean Sea near African coast

HFP/IDP   **HDP-L**

D: HFP/IDP and HDP-L
Examples
- Chad
- Colombia

B: HFP-L only
Examples
- Equatorial Guinea
- Lesotho

E: HFP/IDP and CZTE
Examples
- Adriatic Sea
- Ionian Sea north of the 39th Parallel

F: HFP-L and CZTE
Examples
- Kuwait
- Qatar

G: HFP/IDP, CZTE, and HDP-L
Examples
- Afghanistan
- Sinai Peninsula

C: CZTE only
Examples
- Persian Gulf
- Red Sea

CZTE

SOURCE: Office of the Under Secretary of Defense (Comptroller), 2019, Chapters 10, 17, and 44.

benefits. The figure shows seven areas of overlap, indicating the possible combinations of benefits. As the examples show, each of the seven combinations is possible.

The overlap in eligibility is relevant because the benefit amounts of HDP-L and CZTE can be contingent on receipt of HFP or IDP. This means that creating or terminating a new IDP designation can have a net effect on service members' paychecks that is either larger or smaller than the direct effect of IDP. In particular, as discussed in more detail later, receipt of CZTE can be (but does not have to be) contingent on receipt of IDP, and HDP-L rates are not allowed to be set above $100 per month in IDP-designated zones (the cap is $150 otherwise).

In addition to HDP-L and CZTE, members in specific occupations or who perform certain duties receive special pays in recognition of hazards, although these pays are not necessarily based on location. These pays include demolition duty pay, parachute duty pay, flight deck duty pay, diving duty pay, and toxic fuel/chemical munitions duty pay. Other pays are accession or retention incentives that are not related to risk or location but for which communities frequently exposed to danger may qualify, depending on the needs of the service. For example, enlisted members in special operations may receive an enlistment bonus or reenlistment bonus, and eligible members may also qualify for a critical skills retention bonus. The services have used assignment pays that might, in part, be compensating for hazardous duty. In the past, the Army has used targeted selective reenlistment bonuses that were higher where the probability of deployment to a combat zone was higher. Finally, research since the mid-1990s has found that *military pay*—defined as regular military compensation (the sum of basic pay, the basic allowance for housing, the basic allowance for subsistence, and the tax advantage of receiving allowances tax-free)—is at or above the 70th percentile of the earnings of civilians with characteristics comparable to those of military personnel (OUSD(P&R), 2002, 2012; Hosek et al., 2018). That is, military pay is well above the median pay of comparable civilians. Providing service members with a higher median pay than the median pay for civilians has been deemed necessary to recruit and retain the quantity and quality of personnel the military needs, taking into account the special demands associated with military life (OUSD(P&R), 2002). For example, in addition to being "on call" 24 hours a day and the other daily rigors of military service, service members are often assigned to risky and dangerous locations. Although higher military pay is not targeted to those in higher-risk situations, it is important to recognize in our discussion of HFP and IDP that above-average military pay for all service members already incorporates a compensating differential reflecting the risks and demands of military service.

This report is organized as follows. In Chapter Two, we describe in detail IDP and HFP, including their purpose, eligibility criteria, and benefit amounts. In addition, we describe the IDP designation process as indicated in DoD Instruction (DoDI) 1340.09, *Hazard Pay (HzP) Program* (DoD, 2018), and present tabulations from DMDC data on the incidence of HFP and IDP in recent years. Chapter Two also discusses other risk-based pays related to HFP and IDP, such as CZTE. Chapter Three summarizes the themes that emerged from our interviews of experts, focusing on the four broad questions noted earlier. We then review past studies in Chapter Four, and, in Chapter Five, we review the methodology for awarding danger pay and risk-based pay in a selected set of other organizations, including the U.S. State Department and foreign militaries. Finally, we offer our conclusions in Chapter Six.

CHAPTER TWO

Overview of Hostile Fire Pay, Imminent Danger Pay, and Other Risk-Based Pays

DoD administers its Hazard Pay Program in accordance with U.S. Code, Title 37, Sections 351 and 374 (see 37 U.S.C.; DoD, 2018). That program consists of three types of pay that are designed to address risks inherent in service members' jobs: HFP, IDP, and hazardous duty incentive pay (HDIP). HDIP is not a location-based but rather a duty-based pay. This chapter describes the purpose, eligibility criteria, and benefit levels of these three pays, as well as of HDP-L and CZTE, giving examples of how and when members can earn multiple pays and how those pays interact. We describe the provisions of DoDI 1340.09 that codify the IDP review process and conclude with tabulations of the incidence of IDP and deployments.

Purpose of Hostile Fire Pay and Imminent Danger Pay

Both HFP and IDP were created to recognize risks borne by particular service members in the line of duty, and they evolved over time as the understanding of which risks warranted compensation changed. HFP was established by statute in the Uniformed Services Pay Act of 1963 (Pub. L. 88-132), under U.S. Code, Title 37, Section 310, to recognize frontline combat units for the hazards and hardships they had to endure—specifically, facing hostile fire. This definition of who merited recognition soon shifted from frontline troops to a location-based policy, referred to as *zonal eligibility*: Anyone serving in an HFP-designated location, regardless of the level of risk they faced, merited recognition. The purpose was still limited, however, to recognizing the risks posed by hostile fire.

The Department of Defense Authorization Act of 1984 created IDP by further expanding the types of risk eligible for HFP. Rather than authorizing a completely new pay, the law revised and expanded the authorization for HFP, allowing it to be paid in recognition of "the threat of physical harm or imminent danger on the basis of civil insurrection, civil war, terrorism, or wartime conditions" (Pub. L. 98-94, § 905). Thus, IDP may be understood as a type of HFP that is meant to recognize risks other than just hostile fire.

The "recognition" justification for HFP and IDP is unusual.[1] Appendix A provides a more complete history behind the evolution of risk recognition pay, but it is important to note that most other special pays exist for the purpose of accomplishing particular goals that have measurable outcomes—for example, to fill certain occupations or to reach recruitment quo-

---

[1] This point is elaborated and justified at length in Gould and Horowitz, 2012a.

tas.[2] Recognition, on the other hand, is not associated with a measurable outcome, making it difficult to justify HFP and IDP *a priori* on grounds that these pays achieve a particular goal in a cost-effective way (Gould and Horowitz, 2012a, p. 211).

Hostile Fire Pay and Imminent Danger Pay Eligibility

Eligibility for HFP and IDP is based on meeting one of four criteria:

1. being exposed to a hostile fire event
2. being killed, injured, or wounded in a hostile fire event
3. being on duty in an area in which a hostile fire event occurred that placed the service member in grave danger of physical injury
4. being in a designated area based on the threat of imminent danger (37 U.S.C. § 310(a)(2)).

The first three criteria require a hostile fire event to have occurred and can roughly be thought of as yielding HFP under the original definition, while the last criterion is unrelated to the presence of hostile fire and can be thought of as paying out IDP. Furthermore, the first two can be thought of as event-based eligibility criteria, while the fourth is zonal and the third is a mix. Therefore, each of the criteria requires eligibility to be verified in different ways. Figure 2.1 illustrates the relationships between the different criteria.

For event-based eligibility under the first or second criterion, the hostile fire exposure must be certified by the appropriate on-scene commander or by a death certificate or incident report that establishes hostile fire as the cause of death or injury (DoD, 2018, § 3.2b). These determinations are conclusive and not subject to review (37 U.S.C. § 310(e)). Eligibility under either of these criteria earns a full month of HFP for the month in which the event occurred (37 U.S.C. § 310(b)(2); DoD, 2018, § 3.1.c(2)(a)). In addition, eligibility continues while the service member is hospitalized as a result of the hostile event–related injuries, for up to one year or until the member is returned for assignment, discharged, separated, or retired (DoD, 2018, § 3.1a(3)).[3]

For eligibility under the third criterion, a hostile fire area needs to be defined around the location in which the hostile fire event occurred, delimiting the zone in which service members were in grave danger of physical injury. Hostile fire areas are determined by the relevant service secretary (DoD, 2018, p. 31). While this eligibility category is possible in theory, our interviews with SMEs did not provide evidence that hostile fire areas are used in practice, and, as of June 2019, there were no active hostile fire areas.

Zonal eligibility under the fourth criterion is based solely on location and applies to service members performing duty in an IDP-designated area (DoD, 2018, § 3.3a.(1)). IDP-designated areas are determined by the Assistant Secretary of Defense for Manpower and

---

[2]   For recent studies considering special pays in the context of recruitment and retention incentives, see Hosek, Mattock, and Asch, 2019; Hosek et al., 2017.

[3]   These authorizations are made in U.S. Code, Title 37, Section 372, which additionally provides the relevant service secretary with the power to extend eligibility, in six-month increments, if the service member remains in a medical unit for longer than one year. Note that these provisions are more generous than the statute authorizing HFP, which stipulates payment for up to three months of hospitalization resulting from hostile event–related wounds (37 U.S.C. § 310(a)(3)).

Figure 2.1
Paths to HFP or IDP Eligibility

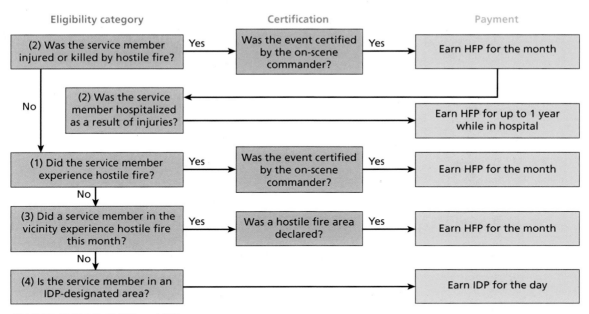

SOURCE: 37 U.S.C. §§ 310 and 372.
NOTE: The numbers in parentheses in the eligibility category boxes correspond to the four criteria identified on p. 6.

Reserve Affairs (ASD(M&RA)) based on geographic demarcations recommended by combatant commanders (CCDRs) and subsequent recommendations from the director of the Joint Staff (DoD, 2018, § 3.3b). Figure 2.2 shows active IDP designations as of June 2019, drawing from DoDI 1340.09.[4] Zonal eligibility for IDP results in a prorated payment for each day spent in the designated area. Service members are not eligible if they are in the designated zone for personal reasons, if they are transiting through, or if they are on leave outside the area.

An important aspect of IDP designations is that the geographic CCDRs are required under DoDI 1340.09 to recommend the smallest geographic area under which the threat exists. Often, country borders are used, but designated areas may sometimes be delimited by particular latitude and longitude coordinates or as circular areas of a certain radius around a central location (see Table B.1 and Figure 2.2 for examples). These stark boundaries mean that service members who leave those regions for a day, even as part of their duties pertaining to the IDP-designated area, will lose a day's worth of IDP. The stipulation also means that eligibility is specified separately for land, air, and water space, and sometimes also for different altitudes of airspace.[5]

To illustrate how the criteria differ in practice and how they interact, we provide three examples. These examples highlight three aspects of HFP and IDP: Only one can be paid at

---

[4] See also Appendix B for a comprehensive list of areas that were actively designated for HFP and IDP at any time since September 11, 2001.

[5] For example, if the airspace is not explicitly designated, then the service member must land in the IDP-designated area in order to be eligible for IDP.

Figure 2.2
IDP-Designated Areas, as of June 2019

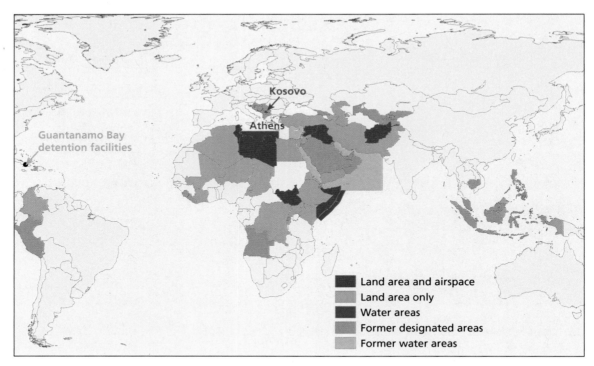

SOURCE: Based on data in Office of the Under Secretary of Defense (Comptroller), 2019, Figure 10-1.

any given time; they have different applications to service members stationed in the United States; and they have different requirements for location-based eligibility.

Example 1: HFP and IDP cannot be paid simultaneously. Suppose a service member has been serving for three months in an IDP-designated location (such as Iraq). He (or she) is then injured by hostile fire at the beginning of his fourth month in that location. He is immediately transported to the United States and hospitalized for three months as a result of those injuries. In this case, the service member had been receiving $225 per month for IDP. He would also be eligible for IDP at the rate of $7.50 per day he spent at the location during the fourth month, up to the time of his injury. But because of his injuries, he becomes eligible for HFP for the entire month, so he earn another $225 (instead of the daily proration for IDP). For the following three months that he spends in the hospital, he would see no interruption in the pay, getting $225 for each of those months. Overall, he would receive $225 each month, but the first three months were paid as IDP and the last four months were paid as HFP.

Example 2: Service members in the United States may earn HFP but not IDP. Service members in the Pentagon or World Trade Center complex on September 11, 2001, were awarded HFP based on their exposure to hostile fire (Kapp, 2005, p. 7). Similarly, HFP was paid to service members who were killed, injured, or wounded in the 2009 attack at Fort Hood, Texas (Hennessy-Fiske, 2015). These examples show that service members located on U.S. soil can earn HFP based on exposure to hostile fire. However, by statute, such service members cannot earn IDP.

Example 3: Zonal eligibility is event-based for HFP but not for IDP. Suppose a service member is on duty in a country not eligible for IDP, but a service member nearby is exposed to hostile fire and the relevant service secretary designates a hostile fire area. In this case, the first member would earn HFP for that month based on her proximity to the hostile fire event. In contrast to this, an IDP designation does not require any event, let alone a hostile fire event, to have occurred. Rather, it requires the threat of an event as defined earlier by the four criteria and deemed significant enough to warrant recognition.

Benefit Amounts for Hostile Fire Pay and Imminent Danger Pay

The pay levels for HFP and IDP are determined by statute. As of September 11, 2001, the pay was $150 per month. As part of the Emergency Wartime Supplemental Appropriations Act, the rate was raised from $150 to $225 for FY 2003 (Pub. L. 108-11, § 1316(a)), and this pay raise was later made permanent.[6] Later, the FY 2012 NDAA mandated that all HFP and IDP paid based on zonal eligibility be prorated to account for the number of days spent in a designated zone (including hostile fire areas) (Pub. L. 112-81, § 616). Therefore, the effective amount became $7.50 per day, up to the monthly maximum of $225 authorized by the law (37 U.S.C. § 310(b)(3)). This proration applies only to zonal eligibility; if a service member is exposed to hostile fire, he or she receives the entire monthly amount for the month in which the event occurred (DoD, 2018, § 3.1c). In practice, however, proration is not applied to HFP, because of an exception provided in the U.S. Code that allows the service secretary, at his or her discretion, to pay the member a non-prorated $225 (37 U.S.C. § 310(b)(2)).

The FY 2008 NDAA consolidated special and incentive pay authorities, establishing in U.S. Code, Title 37, Section 351 an authorization to pay HFP, IDP, and HDIP as versions of hazardous duty pay (Pub. L. 110-181, § 661). As part of this consolidation, the maximal rate for HFP was authorized at $450, and the maximum rate of IDP was authorized at $250 per month (37 U.S.C. §§ 351(b)(1) and 351(b)(3)). DoDI 1340.09 made this consolidation part of DoD policy, transitioning HFP, IDP, and HDIP to be paid under the authorization of Section 351 instead of the previous authorities. As already noted, both HFP and IDP are currently paid in practice at a maximum of $225 per month rather than the maximum authorized by Congress.

Imminent Danger Pay Designation and Review Process as Outlined in DoDI 1340.09

DoDI 1340.09 codifies the recommendation and review process for approving new IDP-designated regions. Section 2.1 identifies the ASD(M&RA) as the authority in administering HFP and IDP (DoD, 2018). This includes setting pay amounts, reviewing the designation

---

[6]   The pay raise was first extended to December 2004 by the FY 2004 NDAA (Pub. L. 108-136, § 619) and then made permanent in the FY 2005 NDAA (Pub. L. 108-375 § 623(a)). Notably, the House of Representatives version of the FY 2004 NDAA provided the $225 rate only for those serving in a combat zone designated for Operation Iraqi Freedom or Operation Enduring Freedom, meaning that all other service members eligible for HFP or IDP would receive $150 per month. This sparked some controversy, as recorded in Congressional Research Service, 2003, and discussed retrospectively in Gould and Horowitz, 2012a, p. 209. For the original House version of the law, see U.S. House of Representatives, 2003, § 622.

requests, and calling for periodic reviews of IDP designations. The DoDI outlines the following roles and responsibilities in the IDP request and review process:

- The director of the Joint Staff evaluates all IDP designation requests and makes recommendations to the ASD(M&RA) regarding all ASD(M&RA)-requested periodic IDP assessments.
- Geographic CCDRs may submit requests for new IDP designations, and they provide written assessments for these requests, as well as for periodic assessments and for any modifications or terminations that they believe are warranted based on changes in the conditions in a given area.
- Service secretaries implement HFP and IDP and evaluate geographic CCDRs' designation requests in coordination with the director of the Joint Staff.
- The Office of the Under Secretary of Defense for Policy reviews the threat assessment provided by the geographic CCDRs, providing a politico-military analysis via the Assistant Secretary of Defense for International Security Affairs, the Assistant Secretary of Defense for Homeland Defense and Americas' Security Affairs, or the Assistant Secretary of Defense for Asian and Pacific Security Affairs, as applicable. The relevant Assistant Secretary of Defense also coordinates with the ASD(M&RA) regarding final details and the effective date of IDP, should it be approved for a particular region.
- The Office of the Deputy Assistant Secretary of Defense for Military Personnel Policy updates the list of IDP-designated areas and transmits to the Office of the Under Secretary of Defense (Comptroller)'s chief financial officer.
- The chief financial officer publishes updates to IDP designations in Volume 7A of DoD 7000.14-R (Office of the Under Secretary of Defense (Comptroller), 2019).

When processing a new request, the ASD(M&RA) receives the CCDR's area assessment, as well as the Joint Staff's recommendation (after consultation with the service secretaries) and the politico-military analysis from the Office of the Under Secretary of Defense for Policy via the relevant Assistant Secretary of Defense. Assuming that the request is approved, the default effective date is the date of the CCDR's most recent area assessment, although other dates can be requested and pay can be backdated prior to the date of designation, subject to availability of funds.

Given this delegation of responsibility, the geographic CCDRs have the sole ability to make new requests. They are also fully responsible for providing relevant information for IDP reviews or updates. This information comes in the form of an area assessment. DoDI 1340.09 (DoD, 2018) provides an area assessment form in Appendix 3A that includes several types of threats and requires that CCDRs indicate those applicable to the region in question, providing both statistics and qualitative narratives as evidence that the threats do indeed place U.S. service members and families in imminent danger.

The main categories of threat are identical to those listed in the statutory definition of imminent danger, with specific itemized subcategories. For example, the potential acts of violence listed in the threat assessment are assassination, homicide, sabotage, kidnapping, aggravated battery, property damage, terrorizing, extortion, rioting, and commandeering/hijacking of a vessel or plane. Other sections ask for particular types of evidence that service members are threatened, such as the presence of travel restrictions or restrictions on dependents. Importantly, all sections ask specifically about risks to uniformed service members and their dependents; the "imminent danger" standard is not sufficient to warrant IDP if it applies only to local citizens, tourists, or other groups.

The DoDI does not prescribe particular types of data or justifications that must be used in an area assessment, and it also does not specify a particular threshold that an area must meet in order to qualify for IDP. Instead, the policy gives the geographic CCDRs leeway in establishing their own procedures and in determining how to fulfill the requirements of the area assessment. For example, the DoDI mandates that geographic CCDRs "establish procedures for the review of designated imminent danger areas . . . to ensure continued designation is warranted" (DoD, 2018, p. 13). Thus, no guidelines for this review process are given in the DoDI other than a reference to the area assessment already described. Similarly, the DoDI states that it is the responsibility of the ASD(M&RA) to prescribe procedures for the periodic review of IDP areas, so it leaves the details, including frequency of the review, unspecified. In 2014, a U.S. Government Accountability Office (GAO) report noted that this lack of guidance on the frequency of reviews dated back to at least 2010 (GAO, 2014).

Other Risk-Related Pays

As mentioned in Chapter One, other pays are related to HFP and IDP. Here, we describe the purpose, eligibility requirements, and benefit levels of HDP-L, CTZE, and HDIP and explain how they relate to HFP and IDP, including any examples highlighting how one pay may be contingent on the other.

Hardship Duty Pay – Location

HDP-L is one of three hardship duty pays, along with hardship duty pay – mission and hardship duty pay – tempo, included in the assignment and special duty pays that are administered in accordance with U.S. Code, Title 37, Section 352 (see DoD, 2019).

HDP-L is designed to provide equity across locations, accounting for substandard living conditions in comparison to the continental United States (DoD, 2019, § 4.4a). As a result, it does not explicitly recognize combat-based risks or danger as defined for HFP and IDP. Rather, substandard living conditions may refer to any aspect of the location that is inferior to the continental United States or poses unusual stress, such as a harsh climate (e.g., in Alaska). For comparison, hardship duty pay – mission recognizes particularly arduous conditions associated with assigned mission or duties performed outside of normal military operations, and hardship duty pay – tempo recognizes extended or excessive amounts of time spent outside of a service member's permanent duty station.

The ASD(M&RA) determines locations eligible for HDP-L. Service members assigned to temporary duty or deployed in eligible locations may begin receiving HDP-L after serving in the location for more than 30 consecutive days; on the 31st day, HDP-L is payable retroactively to the first day the member reported for duty. For those on permanent assignment to the location, eligibility is immediate. HDP-L can be paid in combination with other assignment and special duty pays but cannot be paid concurrently with an assignment incentive pay if that incentive pay is location-based.[7]

---

[7] Assignment incentive pay is meant to encourage service members to volunteer for less-desirable assignments, the reasons for which may include but are not limited to location (DoD, 2019, § 4.1a).

HDP-L is paid in amounts of $50, $100, or $150 depending on location, with the exact amount set by the ASD(M&RA). However, HDP-L is capped at $100 when paid concurrently with HFP or IDP. New IDP designations can therefore result in HDP-L pay being lowered from $150 to $100, as occurred in 2017 when Mali, Niger, and parts of Cameroon were designated for IDP.

Combat Zone Tax Exclusion

CZTE differs from HFP and IDP and from HDP-L in that it is not a payment but a tax benefit. As part of the tax code, it is established by executive order or statute.

Purpose and Eligibility

CZTE is meant to exempt service members from taxation of earnings received when they are in a designated combat zone (DoD, 2010).[8]

A service member is eligible for CZTE if he or she serves either in a designated combat zone, which can be established only by the President (via executive order), or in an equivalent zone established by Congress (via statute). Because a combat zone can be designated only by executive order (26 U.S.C. § 112(c)(2)), the areas designated by Congress are called qualified hazardous duty areas (QHDAs), and military personnel serving there are extended all the tax benefits of combat zones. In practice, the laws establishing each QHDA thus far have specified that receipt of CZTE is conditional on receipt of HFP or IDP. This caveat essentially makes the tax benefits contingent on DoD's determination of the presence of risk; that is, if the area is not deemed to present imminent danger, then CZTE benefits do not apply. However, for combat zones established by executive order, CZTE is a benefit regardless of DoD's determination of risk.

Another route for service members to become eligible for CZTE is by directly supporting a combat zone or QHDA, as designated by the Principal Deputy Under Secretary of Defense for Personnel and Readiness. Neither the Secretary of Defense nor the services can establish a combat zone or QHDA on their own, but they can determine which units are in direct support of one by virtue of their missions. This direct support designation extends the CZTE benefit to those designated units under Treasury Regulation 1.112-1. Direct support eligibility is always contingent on receiving HFP or IDP (26 C.F.R. § 1.112-1(e)(1)).

CZTE benefits continue for up to two years while a service member is hospitalized as a result of wounds, disease, or injury incurred during service in a combat zone or QHDA. Benefits are excludable even if they are paid after the eligibility period is over (for example, after the service member returns to the United States), so long as they are paid for work done during the eligibility period. Similarly, even when service members are hospitalized after the eligibility period, they will be CZTE-eligible if the hospitalization was a result of conditions acquired during the eligibility period (for example, for injuries that are not treated until returning to the United States or for diseases that do not get treated until returning home).[9]

Only military compensation may be excluded, and it is only excludable from federal income taxes (although, as explained later, most states match the federal CZTE either explicitly by law or because of the way they calculate taxable income). This means that Medicare and

---

[8] For a history of the varying justifications for CZTE, see the brief history recounted in Appendix A and the more extensive history recounted in Gould and Horowitz, 2012b.

[9] See the examples written in 26 C.F.R. § 1.112-1.

Social Security taxes still apply. Retirement pay and pensions are not excludable, and neither is income earned as a contractor or from a private firm or other third party, even if it is paid for the same work as is military pay.

As of mid-2019, there were three combat zones: the Arabian Sea, Afghanistan, and parts of the former Yugoslavia. There were also three QHDAs: the Balkans, parts of former Yugoslavia, and the Sinai Peninsula. Since 2001, there have been and still are several areas deemed to be in direct support of a combat zone or QHDA. Figure 2.3 illustrates the regions eligible for CZTE as of June 2019, distinguishing combat zones, QHDAs, and direct support areas. Appendix C provides more information, specifying all areas that had eligibility at any point since September 11, 2001, along with their associated authorizations.

Of the active combat zones as of July 2019, the one for the Arabian Sea was established first, in 1991 during the Gulf War. It now applies to service members deployed in support of operations in Iraq and includes the Persian Gulf, Red Sea, Gulf of Oman, Gulf of Aden, portions of the Arabian Sea, Iraq, Kuwait, Saudi Arabia, Oman, Bahrain, Qatar, and the United Arab Emirates (White House, 1991). Service members in Jordan (since 2003) and Lebanon (since 2015, due to expire in 2020) who are deemed to be in direct support of the Arabian combat zone also receive the CZTE benefit, and those in Turkey were included from 2003 to 2005.

Figure 2.3
Areas with CZTE Eligibility, as of June 2019

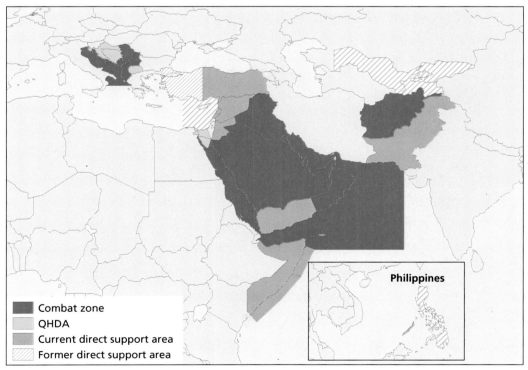

SOURCE: Based on data in Office of the Under Secretary of Defense (Comptroller), 2019, Figure 10-1.

The combat zone in the former Yugoslavia was established in 1999 and includes Serbia, Kosovo,[10] Montenegro, Albania, the Adriatic Sea, and the part of the Ionian Sea north of the 39th Parallel. One month later, Congress designated the same areas part of a QHDA (Pub. L. 106-21).

The Afghanistan combat zone was established in 2001 to include Afghani land and airspace (White House, 2001). It also applied to service members deployed as part of Operation Enduring Freedom, regardless of location. This has so far included certain service members in the Philippines (from 2002 to 2015) and Somali air and water space (since 2007). Eligibility based on direct support criteria has also been extended to certain units: Djibouti (since 2002), Jordan (since 2001), Kyrgyzstan (from 2001 to 2014), Pakistan (since 2001), and Somalia (since 2004); those at Incirlik Air Base in Turkey were included from September 2001 to December 2005.

The Balkans QHDA was formed in 1995, covering Bosnia-Herzegovina, Croatia, and Macedonia (Pub. L. 104-117). The QHDA is still in effect, but the CZTE benefits ended in 2007 when the IDP designation was terminated in those countries. Were these areas to be designated for IDP at some point in the future, they would automatically qualify for CZTE benefits once again.

The Sinai Peninsula QHDA was established as part of legislation first introduced as the Tax Cuts and Jobs Act of 2017, providing CZTE retroactively to June 2015 (Pub. L. 115-97).

Benefit Amount

Because it is a tax benefit, the pecuniary value of CZTE varies based on service member income. For eligible enlisted service members, all military pay is tax-exempt. For officers, the maximum amount that may be excluded is equal to the basic pay for the most-senior enlisted service member (regardless of whether that person was also earning the CZTE benefit at the time) plus the amount of HFP or IDP payable for the qualifying month.[11] The benefits are guaranteed only at the federal level; state and local tax treatment of military compensation varies. At the state level, CZTE is guaranteed in most (and perhaps all) states, sometimes by default because the state exempts all military pay or military pay earned outside of the state,[12] sometimes by extending federal CZTE benefits in state law, and sometimes because the state considers taxable income to be defined in accordance with federal guidelines.[13]

Calculating the value of CZTE requires an estimate of how much individual service members would have been paid in the absence of the benefit. As described in Pleeter et al., 2012, the authors worked with the Treasury Department to estimate the range of benefits in 2009. The average eligible service member had a CZTE benefit of $5,990; the median was $4,660; and total benefits topped $3.6 billion, with a range from $280 (1st percentile) to

---

[10] At the time, Kosovo was part of Serbia but is now an independent state and is included in the combat zone.

[11] See Office of the Under Secretary of Defense (Comptroller), 2019, Chapter 44, subparagraph 440202.A.2. For example, as of January 1, 2019, the basic pay for the Sergeant Major of the Army is $8,578.50. This, plus any HFP or IDP, would be the maximum amount that an Army officer could exclude from income tax. For rates of basic pay, see Office of the Under Secretary of Defense (Comptroller), 2019, Tables 1-7 through 1-10.

[12] For example, as of 2006, Arizona exempts all active-duty military pay from state income tax; see Arizona House Bill 2795 (Arizona House of Representatives, 2006).

[13] For example, Ohio (among several other states) defines *adjusted gross income* in accordance with federal law as laid out in the Internal Revenue Code. See Ohio Revised Code 5747.01. For examples of how states treat military income, with links to each state and territory tax bureau, see Military.com, undated.

$22,430 (99th percentile). Because CZTE benefits allow senior officers to qualify for the federal earned income tax credit, which is normally for low-wage earners, the benefits of CZTE can skew toward high earners. For example, a service member in the O-6 pay grade may earn more in earned income tax credit than someone in the E-4 pay grade does.

Hazardous Duty Incentive Pay

Along with HFP and IDP, HDIP is the third component of DoD's Hazard Pay Program and is also described in DoDI 1340.09.

Purpose and Eligibility

As the name suggests, HDIP provides an incentive to meet critical manpower needs—specifically, particular designated duties that risk physical injury or are inherently dangerous. To establish that a duty is indeed critical, in the designation request to the ASD(M&RA), a service secretary must justify why readiness requirements mandate that the duty be performed on at least a monthly basis.

Designations for HDIP are made by the ASD(M&RA) based on requests issued by service secretaries. An individual becomes eligible for HDIP by virtue of performing a designated duty under competent orders. Eligibility is also contingent on a service member having completed qualifying training to perform the duty (or currently being in such training). Service members receive HDIP from the day they begin eligible duty until either the termination of the duty or the termination of the order to perform it. A service member may receive HDIP for up to three hazardous duties at the same time, and officers may not receive HDIP if they are simultaneously receiving an incentive pay for the same skill. See Table D.1 in Appendix D for a list of designated duties (as of June 2019) with associated training requirements.

Because it is part of DoD's Hazard Pay Program, HDIP will continue to be paid through any period of hospitalization so long as the illness or injury occurs while serving in a combat zone or hostile fire area or during exposure to a hostile fire event (DoD, 2018, § 3.1.a.(3)).[14] In this way, HDIP is similar to HFP and IDP. Outside of combat zone–related hospitalization, HDIP can be paid continuously for up to six months if a service member is temporarily (but not permanently) unable to perform the duty because he or she is sick, injured, or on authorized leave.

As in the section on HFP and IDP eligibility, we offer examples to illustrate how HDIP differs from those pays and how they all interact.

Example 1: Receiving HDIP without HFP or IDP. Suppose that a service member earns HDIP for laboratory duty involving live dangerous viruses or bacteria. She is assigned to a laboratory in the continental United States. Therefore, she does not earn IDP or—barring an extreme event, such as a hostage situation at the lab—HFP.

Example 2: Receiving HDIP and HFP or IDP. Suppose that a service member is assigned to flight deck duty, and his ship is located in the part of the Somali Basin designated for IDP (see Table B.1). If he is assigned on deck for flight operations on at least four days in that month

---

[14] The continuation of hazard pay is authorized by U.S. Code, Title 37, Section 372 and extends to other types of incentive pays, bonuses, and the daily incidental expense allowance. Collectively, the continuation of these pays is guaranteed by the Pay and Allowance Continuation Program; see Office of the Under Secretary of Defense (Comptroller), 2019, Chapter 13, para. 1302, and especially subpara. 130203.A.

(see Table D.1), he would earn $150 of flight deck HDIP that month in addition to $7.50 in IDP for each day the ship spends in the Somali Basin.

Example 3: HDIP contingent on HFP or IDP or on presence in a combat zone. Suppose that a service member is eligible for HDIP for duty involving handling chemical munitions, and he is assigned to perform this duty in Iraq. He would earn both HDIP and IDP, as in Example 2. If he were to be injured by hostile fire (unrelated to the performance of the hazardous duty) and transported to a hospital to recover from those injuries, then he would continue to receive HDIP during hospitalization.

Similarly, if a service member were serving in Kuwait and contracted an illness in the line of duty, HDIP would be continued during the subsequent hospitalization. This is because Kuwait is part of a combat zone, even though it is no longer designated for IDP.

Benefit Amount

The amount of HDIP depends on the particular duty, varying from $150 to $250 per month (see Appendix D for rates). The payment is prorated according to the number of eligible days in a given month, and the services have discretion in determining payments. When a duty is designated for HDIP, the services may offer the payment but do not have to, or they may set payment at an amount below the maximum authorized by the ASD(M&RA).

HDIP is authorized under U.S. Code, Title 37, Subchapter 1 (Section 301) or Subchapter 2 (Section 351), although, at any given time, an individual could receive it under only one or the other, not both. In January 2018, DoD policy changed to begin paying HDIP as part of the Hazard Pay Program under Section 351 rather than Section 301, and HDIP may no longer be paid under Section 301 (DoD, 2018).

The policy change to pay HDIP based on Section 351 has implications for HDIP rates. A major difference between the two authorizations is that Section 351 stipulates that HDIP rates may not vary by pay grade (37 U.S.C. § 351(g)). Flying duty pay and diving duty pay, which previously varied by pay grade and years of experience, were therefore revised to standardize the rates (see Table D.1).[15] In addition, diving duty pay is now paid as a version of HDIP, but previously it was a separate incentive pay (authorized under 37 U.S.C. § 304).[16] The policy change also resulted in the elimination of a separate pay for air weapon controller crew members; they would now be paid flying duty pay as part of an air crew. Finally, service members may earn HDIP for up to three duties at the same time; under Section 301, the maximum was two duties. The rates and stipulations listed in Table D.1 in Appendix D reflect these changes.

Incidence of Risk-Related Pays and Deployment

We conclude this chapter with tabulations of cumulative deployments and the payout of risk-related pays from FY 2012 to FY 2018. Our tabulations focus on the Army—the service with the most cumulative deployments—and on members who entered active-duty service after October 2012 so that we can observe their entire service through 2018. The data used are the

---

[15] For the previous rates varying by grade, see Office of the Under Secretary of Defense (Comptroller), 2019, Chapter 22 (for flying duty) and Chapter 11 (for diving duty).

[16] DoDI 1340.09 stipulates that service members may not receive diving duty pay under Section 304 if they are also receiving HDIP under Section 351 (DoD, 2018, § 3.1.b.(1)(a)).

DMDC Contingency Tracking System (CTS) data for deployments and the DMDC Active Duty Pay File. For each month in active duty, the data record special pays that were paid out in that month (including type of pay, location associated with the pay, and dollar amount), whether the person was eligible for CZTE during the month (and location meriting eligibility), and location of deployment in the CTS, if applicable.

Figure 2.4 shows cumulative person-months of Army members and affirms the expected patterns, with several notable features. The solid black line shows the cumulative months of deployments in the CTS data, while the dotted line shows deployments that would merit IDP; Bahrain, Kuwait, and Qatar merited IDP until 2014. The orange line shows cumulative person-months of HFP and IDP payouts from the pay files, and it closely tracks the dotted line, as expected. It is slightly higher than the dotted line, reflecting the presence of troops in nondeployed IDP locations. Overall, the orange line shows that the payout of HFP and IDP to Army members tracks their deployments, as measured in the CTS data.

The blue line in Figure 2.4 shows the cumulative months of CZTE eligibility. This line also closely tracks overall months of deployment because the CTS deployment locations with the vast majority of troops—including Bahrain, Kuwait, and Qatar—are in combat zones or direct support areas and therefore merit CZTE benefits. We find that there are more people earning CZTE than earning IDP, which is expected because not all CTZE-eligible locations are eligible for IDP. The green line shows months of HDP-L payouts. This is the largest of all the categories in Figure 2.4 because HDP-L is paid in many locations for many reasons, including risk. Therefore, we would not expect HDP-L to track any of the other variables being plotted, and it is not possible to verify that the amounts of HDP-L are as expected, because we cannot calculate how many individuals were stationed in each eligible location.

Figure 2.4
Cumulative Months in Deployment and Months of Risk-Related Benefits for Active-Duty U.S. Army Personnel with Service Commencing After October 2012

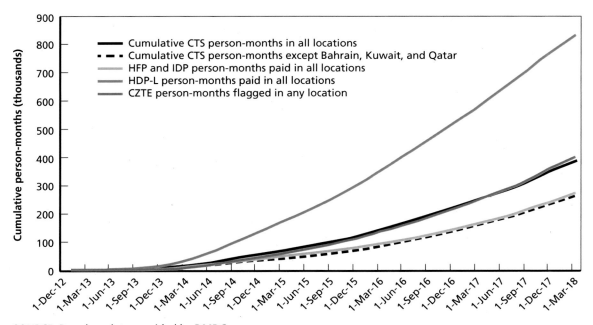

SOURCE: Based on data provided by DMDC.

We also investigated whether we could identify gaps whereby service members were deployed but did not receive HFP or IDP and in which locations these gaps occurred. Such gaps could provide an indication of whether the current location-based methodology for awarding HFP and IDP was inadequate by not providing a special pay to service members deployed and potentially in harm's way.

Unfortunately, available data from DMDC are not suitable for this type of analysis, for several reasons. First, the pay files indicate when HFP or IDP was paid, not when that payment was earned. Because of delays in processing, including the payment in one month for multiple months of eligible service, HFP and IDP may not be paid in the month that they were earned. Consequently, when examining the data, we would observe false positives showing that members were deployed but not receiving HFP or IDP. Second, the CTS file captures deployments in support of operations for the Global War on Terrorism. A service member might be overseas and in a threatening environment but not show up as a deployment in this file if his or her activity is not considered a deployment or a deployment in support of Global War on Terrorism operations. Finally, although the CTS data indicate where the member was deployed and the pay file indicates the location (at least as of 2017), in many cases in which the member earned HFP or IDP, the location is listed as "unknown." Consequently, we are unable to fully identify locations where individuals are deployed but not receiving HFP or IDP; conversely, even in locations where we know that members are eligible for HFP or IDP, we are unable to fully account for all payments and deployments. For example, we find that, in Afghanistan—where all deployed members should receive IDP—the CTS data indicate 188,500 person-months of deployment, but the Active Duty Pay File data indicate only 100,323 person-months of IDP payouts, because some who deployed to Afghanistan have "unknown" as the location associated with their IDP payouts.

The implication is that DMDC would need to develop additional data sources to enable DoD to assess whether there are gaps between the payment of HFP or IDP and the deployment of service members. Appendix E describes the DMDC data in more detail and shows tabulations by country.

CHAPTER THREE

Major Themes from Interviews with Subject-Matter Experts

We conducted 20 interviews with senior officers and civilians in positions that were relevant to IDP recommendation, review, or implementation. The interviewees were drawn from all levels of the IDP review process, as outlined in DoDI 1340.09 (DoD, 2018): the services, the geographic COCOMs, the Joint Staff, and relevant DoD offices. We also talked to individuals from nongeographic commands, the National Guard, and particular installations that were deemed relevant to some of the concerns raised about HFP and IDP. Personnel from the project sponsor (the Office of the Deputy Assistant Secretary of Defense for Military Personnel Policy) were responsible for contacting the interviewees and scheduling the interviews and were instrumental in defining the scope of our interview list and in recommending particular individuals who would provide the greatest insight into IDP.

Using a semi-structured interview protocol, we sought participants' feedback on HFP and IDP, with special focus on the four topics that address the elements of the report, as required in the FY 2019 NDAA (Pub. L. 115-232, § 612(b)):

1. What is the current methodology for making an HFP or IDP designation or recommendation? (addressing § 612(b)(2))
2. Is the current IDP process effective? Does it meet the needs of service members, including special operations forces? (addressing § 612(b)(3)(A))
3. Is geographic eligibility the best method for awarding IDP, or would a different approach, particularly one based on deployment or operations, be more effective and efficient? (addressing §§ 612(b)(3)(B) and 612(b)(3)(C))
4. What are the difficulties in implementing the current system? (addressing § 612(b)(1))

The semi-structured nature of our interviews allowed us to explore the unique experiences and observations of our interviewees, in both their current and past positions. Although perspectives and observations differed among the experts, several themes emerged from the interviews based on what interviewees identified as key issues and on what they revealed, according to their experiences and interactions, as being of value to service members. We summarize these themes here.

The Current Review and Recommendation Process for Imminent Danger Pay Hews Closely to DoD Guidance and Is Considered to Be Broadly Effective

Interview participants were in unanimous agreement that DoDI 1340.09 (DoD, 2018) is the policy source document for performing IDP area assessments. In practice, the process works largely as prescribed by the DoDI. To fill out the assessment form, geographic COCOMs work with the geographic J-1, J-2, J-3, and J-5 directorates to compile justifications based on intelligence reports, crime statistics, and other information. The recommendation packet is sent to the J-1 at the Pentagon, and it is circulated to the individual services for review and recommendation.

In general, interviewees indicated that the list of threats for defining *imminent danger* were reasonable and clear. Among those for whom it was relevant, the area assessment form was considered to be clear and unambiguous, as were the necessary justifications. When asked whether the request and review process allowed for proactive consideration and acknowledgement of threats—rather than responding to threats after an adverse event occurs—the participants felt that the answer was generally yes.

Interviewees were also in unanimous agreement that the current process "works." Comments we heard included that "it's hard to find something as effective and not burdensome," "it ain't broke," and "I'm not aware of people falling through the cracks." Overall, interviewees did not mention that they had heard complaints about IDP being paid late or inconsistently. Interviewees felt that it was rare for a service member to have IDP missing from a paycheck, and, in such cases, the omission could be resolved by filing the appropriate paperwork and having the pay issued retroactively. This included any situation in which special operations forces may have missed pay because of the classified nature of a mission.

In short, interviewees said that the system operated well to provide pay to service members exposed to the threat of danger while also remaining relatively easy to administer.[1]

The Current Review and Recommendation Process for Imminent Danger Pay Is Not Perfect, and There Are Areas for Improvement

Multiple interviewees mentioned areas in which the IDP process or the IDP benefit itself were less than optimal. These areas fall into five broad areas: agility and timeliness, recertification and perceived accountability, fairness and perceived arbitrariness in the definition of *threat*, adequacy of the dollar amount of IDP and its connection to recruiting and retention, and timely updates and information-sharing. As one interviewee put it, "The process is effective but not efficient." It is important to note that not all interviewees mentioned these areas for improvement, and, in some cases, other interviewees offered counterarguments for why these areas do not require improvement.

Agility and Timeliness
The first area has to do with the agility and timeliness of the process. Although the steps of the process are clearly laid out in DoDI 1340.09, they can take a long time, and some interview-

---

[1]   The term *easy* should be considered as a relative term because, as one interviewee put it, the current system "is ponderous on a good day."

ees indicated that the lengthy review process is one reason for retroactive IDP designations, as occurred in Mali, Niger, and parts of Cameroon in 2017. A few individuals reported that packets had been under review for multiple years, and they still did not know when a determination would be made; it is unclear to what extent this is an exception rather than the norm. Furthermore, as one interviewee noted, with a rapidly changing environment, the factors that led to a location being considered threatening at one point could no longer be relevant at a later point and, importantly, by the time an IDP package was approved. Put differently, some interviewees argued that the review process is not agile enough to respond to adverse events in a timely way. They stated that an indeterminately long review cycle could have implications for troop morale. As an example, a few people mentioned the recent case of Mali, Niger, and parts of Cameroon, which were recently designated for IDP. Five months elapsed between the events that triggered the request—the death of four troops in Niger on October 4, 2017—until the final approval in March 2018. The upshot was that troops in or near Niger during that time felt that they were being placed in obviously imminent danger without the requisite compensation. Although IDP was eventually paid retroactively (back to June 7, 2017) (Office of the Under Secretary of Defense (Comptroller), 2019, Chapter 10), it did not change the perception that morale was affected in the interim. (This five-month turnaround time was in fact shorter than what interviewees generally indicated was the norm.)

Several people noted that part of the reason for the long process is that "IDP is a very political decision." In particular, declaring a country as posing imminent danger to U.S. troops can be politically problematic if that country is also a member of the North Atlantic Treaty Organization or is a U.S. ally. The result is that doing due diligence in setting IDP requires input from a large number of stakeholders, such as the COCOMs, the services, the Joint Staff, OUSD(P&R), the General Counsel, and the Office of the Under Secretary of Defense for Policy.

Recertification and Perceived Accountability
One theme that arose in nearly every interview had to do with the frequency of reviews for the locations already designated for IDP and whether periodic review of locations should be institutionalized or done on an ad hoc basis. The interviewees provided mixed views about whether reviews should occur with a predefined frequency. On the one hand, some argued that IDP originates "on the ground" and from the COCOMs and that CCDRs can request removal of a country, as was recently done for Greece. Thus, the current process already allows for ongoing review. Furthermore, we learned that a review cycle is currently being completed, and the most recent prior review was in 2014.

That said, some interviewees questioned whether the "COCOMs are held to account." One interviewee made the point that incentives are misaligned in the designation process. The CCDRs can request a designation, but the cost is borne by the services; that is, "COCOMs designate but services pay." Consequently, according to this interviewee, the COCOMs have little incentive to remove a designation. Others wondered about the "optics" of having some locations qualifying for IDP for 34 years or of having locations designated for IDP but also allowing dependents, as in Colombia. One interviewee gave the anecdote of cruise ships being allowed to dock in a country designated for IDP. Although interviewees noted that these cases could have legitimate justifications, the lack of a predetermined periodic review or, alternatively, a sunset provision whereby IDP designation would end unless a new review renewed the designation could give credence to the optics that these cases reflected a lack of accountability.

Fairness and Perceived Arbitrariness in the Definition of *Threat*

Designation of IDP requires a line to be drawn to delimit IDP-designated regions, thereby creating "haves" and "have nots." For the "have nots," perceived inequities occur when service members cross the line, such as when their assigned duties take them over the border; they can lose IDP eligibility even though the enemy and the resulting threat still exist. A few interviewees suggested the creation of "buffer zones" extending past country borders to allow for eligibility to spill over the region of concern or creating multiple concentric zones so that IDP amounts taper off gradually rather than dropping to zero just over a border. Any alternative, however, can only mitigate and not completely solve the border problem. Furthermore, a few people noted that, in a dynamic threat environment, managing such buffer zones could be administratively costly.

Several interviewees said that the designation process will tend to define IDP geographic locations rather broadly and keep a designation for a long time to minimize "have nots." As one interviewee said, "We don't want to disadvantage anyone, so IDP designations remain around for years." As this statement suggests, the result of this approach is that several interviewees felt that IDP designations last too long and can outlive the original justification. Furthermore, they noted that actual threats and levels of danger in IDP-eligible regions can vary widely.[2] Such differences can create the perception of unfairness and undermine any morale-boosting value of IDP in areas with the most-severe danger (e.g., Iraq and Afghanistan). Interview participants largely agreed that the 2014 worldwide review was effective at resolving some of these discrepancies because it resulted in the removal of several countries (e.g., Kuwait and Bahrain) for which there was nearly unanimous agreement on the lack of danger to U.S. troops. Some interviewees felt that removing IDP designation should be made a matter of policy—for example, by having IDP automatically expire after a certain number of years unless it is reviewed and recertified, as discussed earlier.

A few interviewees mentioned that some inequities in IDP eligibility could be mitigated by instituting more-narrow geographic designations. Country-level designations are most common but are not used in every situation and are not mandatory; indeed, DoDI 1340.09 requires that CCDRs' requests be "limited to the smallest geographical area (e.g., portions of countries or seas) in which the danger or threat exists" (DoD, 2018, p. 12). These interviewees felt that defining IDP-eligible areas more narrowly than entire countries is currently merited because the threat level is often sufficiently varied, even within country borders. As a counterpoint, other interviewees pointed out that narrower definitions can pose administrative burdens because it is difficult to track service members' precise locations within a country at all points in time. As one interviewee stated, "it would be impossible to manage this surgically."

Several interviewees said that the varying levels of threat within and across IDP-designated areas made the designations appear arbitrary. As one interviewee said, "the IDP assessment is weak" because IDP is not based on the severity of the threat.

---

[2]   This point was made in the Eleventh QRMC (specifically, Pleeter et al., 2012), which quoted an opinion article in the *Washington Post* that appeared in March 2011 entitled "I Didn't Deserve My Combat Pay." The author of the opinion piece described the conditions surrounding his deployment in Iraq as being safe and the living conditions "plush."

Adequacy of the Dollar Amount of Imminent Danger Pay and Its Connection to Recruiting and Retention

The IDP amount is $225 per month, prorated, as discussed in Chapter Two. Several interviewees noted that the IDP amount is less than the Family Separation Allowance ($250 per month). As one interviewee said, "The standards are skewed. We pay more for family separation than for IDP. How do you explain that to someone? Being away from their kid is worth more than being in danger?" An additional point made by a different interviewee is that the value of IDP has eroded because it has not been increased in a long time.

Related to the level of the IDP rate is its relationship to readiness. One interviewee remarked that there is no connection between the combat pays and recruiting and retention. Furthermore, this individual noted that hostile environments are already "priced into basic pay," stating that paying IDP is like "paying a fireman to fight a fire." That said, the interviewee recognized the other school of thought that those exposed to greater risk should be compensated but still noted that the recognition of qualifying for combat awards may be worth more to service members than the payment itself does.

Timely Updates and Information-Sharing

Another common theme among interviewees was the greater need for timely updates and the sharing of information about the progress of IDP packages and the list of designated locations. Several interviewees said that they had no visibility on where the IDP package was in the review process or why it was taking so long. Consequently, they were unable to provide information to members and those with boots on the ground about when they might expect IDP to appear in their paychecks, if the designations were approved. As one interviewee stated, "We do a good job, but we need to communicate and have visibility about what's going on." It is possible that this concern reflects a lack of understanding about the large number of stakeholders involved in the process.

A couple of interviewees also pointed out that the available lists of designated countries were not up to date and were sometimes inconsistent. The most common lists are from Chapter 10 of the DoD Financial Management Regulation 7000.14-R, Volume 7a (Office of the Under Secretary of Defense (Comptroller), 2019) and from the Defense Finance and Accounting Service website. These interviewees stated that IDP implementation might be improved by having a single source of up-to-date information—perhaps a separate internal website available to those responsible for such implementation.

The Review and Recommendation Process for Imminent Danger Pay Raises Unique Issues for Special Operations Forces

The interviews generally indicated that special operations forces present special concerns regarding IDP. Interviewees most often indicated that, if anyone were to be liable to fall through the cracks and miss out on earning a deserved pay when operating in a nondesignated region, it would be special operations forces. There are several reasons for this. By the nature of their duties, special operations forces face danger on most (perhaps even all) of the missions they perform, yet they are not present in a region in particularly large numbers or for long periods of time. Furthermore, the geographic eligibility criteria for IDP mean that any pay extended to a region would apply not only to special operations forces but to any service member who enters the zone. The justification for the pay must be based on risks to the average service member, which may be different from the risks to special operations forces.

According to several interviewees, a further hindrance to paying IDP based on the (potential) presence of special operations in a region is that such missions may be classified, while the pay records and the IDP designation list are not. It would not be feasible to pay IDP in locations where the presence of U.S. troops is a matter of national security but the location is classified.

Some interviewees acknowledged that these characteristics of special operations forces create possible limitations for the IDP process. Yet some downplayed this potential misalignment of IDP with special operations' unique dangers, noting that these forces earn special and incentive pays that are designed to compensate them for these precise aspects of their jobs. Besides compensating for risks, those pays avoid the sensitive issue of classified missions because they are automatically paid for particular activities, regardless of location or classification level.

To inquire about how special operations forces' needs are addressed, we asked SMEs about the role of U.S. Special Operations Command (SOCOM) in the recommendation and review process. DoDI 1340.09 gives the right and responsibility to geographic COCOMs and their CCDRs to make new requests for IDP designations and to provide area assessments for these requests or for any periodic reviews (DoD, 2018). Nongeographic commands are not included as part of the formal process as detailed in the DoDI, but they may have information regarding the risks borne by particular service members that could be relevant for an area assessment. Therefore, we sought to understand whether and how nongeographic commands are included in the review process.

In terms of SOCOM's role in the process, interviewees provided mixed responses. In some cases, interviewees reported that the relevant special operations regional command was consulted and, in some cases, closely involved; in others, they felt that the geographic COCOM was sufficiently aware of any relevant information that would have been supplied by SOCOM. And although SOCOM and other nongeographic commands cannot officially originate a designation package (per DoDI 1340.09), some interviewees indicated that they could do so in practice if they routed the request through the relevant geographic COCOM. Finally, interviewees said that all packages relevant to special operations forces eventually "went through."

Nevertheless, some people felt that SOCOM should be given more ownership of the IDP process, especially the origination of IDP packages. They argued that doing so would allow valuable information to be provided during the review and recommendation process that was not currently provided regularly. Consequently, they felt that leaving nongeographic commands out of the official review process meant that such information would be received only on an ad hoc basis. One interviewee stated that the geographic COCOMs would still be important to the IDP process even if SOCOM took a more active role because geographic-related factors unrelated to special operations forces were also relevant to the IDP determination process.

A Deployment or Mission-Based Methodology Would Be Less Desirable

Nearly all interviewers agreed that location-based eligibility for IDP was preferable to alternatives, such as mission-based, deployment-based, or duty-based eligibility. There were four rationales for this conclusion, and they were based on the nature of the definition of *threat*, the fairness and equity of the eligibility rules, the feasibility and administrative burden of different criteria, and the lack of need for an alternative approach.

Several interviewees stated that location is the key element in defining *threat*. As stated by one interviewee, "threat is inherently related to geography." The conclusion these SMEs drew is that it will always make sense to include geography in the eligibility criteria for IDP.

The fairness rationale is the argument that basing IDP on operations, missions, or deployments will lead to inequitable results, given the way the military defines these terms. Interviewees explained to us that operations are not defined in terms of geography, so troops may be supporting an operation yet be stationed far away from danger, including in the United States. Consequently, paying IDP by operation would cast too wide of a net. Similarly, deployment-based standards would not align properly with locations presenting imminent danger. In many cases, it will be too narrow a criterion. Deployment refers to a particular mission status that applies to many, but not all, service members stationed in IDP-designated areas. For example, deployment-based eligibility would exclude most service members in Africa from earning IDP, apart from those in Djibouti, who may be supporting operations in Iraq or Afghanistan. (See Appendix E and discussion in Chapter Two for evidence of how many service members earn IDP but are not deployed.) And, in some cases, deployment may apply to areas that are not dangerous. The example of Kuwait arose repeatedly during the interviews: Service members in Kuwait are considered deployed (moreover, they are serving in a combat zone, so they are eligible for CZTE), but Kuwait was removed from the IDP list in 2014 after a worldwide review determined that it and several other countries (such as nearby Bahrain) did not pose imminent dangers to U.S. troops. Multiple interviewees noted that the presence of military dependents in Kuwait served as strong evidence that Kuwait did not merit IDP, and others noted that it was simply unfair to leave Kuwait on the list after one compared the situation in Kuwait with that in Iraq and Afghanistan.

The feasibility rationale is that geographic eligibility is the most straightforward approach to implement while still providing HFP or IDP to those exposed to the threat of danger. Interviewees argued that other standards would be extremely burdensome and potentially impossible to implement, at least in a timely or relevant manner. For example, a few interviewees noted that a mission-based criterion may technically allow for a more tailored designation of IDP, perhaps in conjunction with geography so that the pay is earned based on a combination of mission type and mission location. But this approach works only in theory. In addition to requiring a list of criteria that missions must meet, the approach would require that review and approval of IDP be allowed at a much lower level than current policy dictates, perhaps at the level of the on-scene commander. This would introduce a degree of subjectivity that could make it difficult to ensure consistency and fairness and could lead to results that were insensitive to political considerations. Some people further indicated that mission tempo could interfere with paying IDP, because missions may occur on short notice and leave little time to consider whether IDP is warranted.

The fourth rationale that we heard for rejecting mission-, deployment-, and duty-based approaches is that there is no need for any alternative. While IDP compensates for threat, HFP compensates for exposure to an actual event and can be originated by commanders in the field. Consequently, some interviewees argued that DoD already has a mechanism built into the HFP and IDP process to recognize exposure to danger that is not location-based.

Interviewees Offered Suggestions for Ways to Improve the Current Review and Recommendation Process for Imminent Danger Pay

While interviewees endorsed keeping the current location-based methodology for awarding IDP, they offered several suggestions for changing the benefit level and structure, the current IDP review process, or the implementation process to address the potential areas for improvement discussed earlier.

First, several interviewees had suggestions for how to change the IDP benefit to more closely connect it to risk. One suggestion we heard was to reform IDP so that it is always event-based—basically HFP—and to define a location around that event. The dollar amount under this suggested reform would be higher than the current amount (and above the amount of the Family Separation Allowance) when the event occurs and then taper off over time unless another event occurs within a prespecified period. Another reform suggestion was to create a tiered set of IDP rates based on severity of threat to recognize that severity is not equal across locations or over time in the same location. Finally, interviewees suggested that the IDP rate be increased to restore its value relative to when it was created and to ensure that it exceeds the Family Separation Allowance, although some SMEs acknowledged that such changes could increase costs.

Second, some interviewees recommended that IDP designations be reviewed on a predetermined regular basis even if that basis is, as one interviewee said, "every ten years." A few recommended that designations include a sunset provision in which designation expires after a certain number of years unless recertified.

Third, several interviewees requested improved communication and information that would allow tracking an IDP package through the IDP process. Also requested was an internal resource, such as a webpage, that would provide a single up-to-date source of information about current IDP-designated locations.

Fourth, one interviewee suggested that the criteria for designation in DoDI 1340.09 include health risks. In particular, the interviewee mentioned risk to service members of exposure to Ebola as an example. The current DoD guidance allows health risk to be included in the IDP package in the section on "other factors," but it does not explicitly request information about health risk to service members.

Finally, some interviewees requested that SOCOM be given a more active role in the IDP designation process, especially the origination and tracking of packages. That said, as mentioned earlier, other interviewees did not see the need for this or expressed concern that geographic COCOMs were still important to the process, given that factors unrelated to special operations forces were also relevant to the IDP determination process.

Interviewees Also Mentioned Hostile Fire Pay and the Combat Zone Tax Exclusion and Their Relationship to Imminent Danger Pay

Interviews also touched on the relationship of IDP to other combat-related benefits. Two general themes emerged: the role of HFP in countering certain limitations of IDP and the importance of IDP to CZTE.

Many interviewees noted that HFP serves as a backstop for IDP. In areas where IDP is not designated, or not yet designated, HFP was viewed as the catchment should a hostile event occur. As mentioned earlier, HFP addresses concerns that IDP is not mission- or deployment-based, and interviewees mentioned that HFP could be viewed as a solution to the problems regarding special operations forces, because HFP can be paid anywhere in the world, even during classified missions. An added benefit of HFP mentioned in the interviews is that it is certified on site and so is paid in a timely manner.

But interviewees also acknowledged that, because it is event-based, HFP is not a perfect backup policy for IDP. When troops are in a situation that might otherwise be deemed to present imminent danger on par with IDP-designated regions but no hostile event occurs, they go home with no additional pay recognizing that danger.

With regard to the CZTE benefit, several interviewees remarked that an important but often-overlooked value of IDP is its role as a prerequisite for certain CZTE eligibility. In QHDAs and direct support areas (but not combat zones), income is tax-exempt only if the service member is also earning IDP. In such cases, the value of the tax exemption is likely higher than the value of IDP itself. As one interviewee stated, "the real money is in CZTE."

This link between IDP and CZTE can make it difficult to remove the designation from certain IDP-eligible countries, according to the interviewees. It also creates inequities in CZTE. For example, the elimination of IDP in Bosnia, Croatia, and Macedonia in 2007 resulted in the loss of CZTE because these countries are part of a QHDA. But because Bahrain, Kuwait, Serbia, and Qatar are part of combat zones designated by executive order, the 2014 removal of the IDP designation in those countries did not affect CZTE. As a result, troops in Serbia still earn CZTE, while those next door in Croatia do not, despite the fact that neither country is currently deemed to present imminent danger. In a further nuance in the law, someone in Croatia *could* receive CZTE if he or she were in a hostile fire situation, because HFP would reactivate CZTE eligibility.

CHAPTER FOUR

Insights from Past Studies

Several previous studies have also examined the adequacy of risk-related pay, including HFP and IDP. In this chapter, we briefly review the conclusions and recommendations of studies conducted since 2001. Some of the themes that emerged in our SME interviews (see Chapter Three) were noted in past studies.

Koopman and Hattiangadi Report on Deployment Pay (2001)

Koopman and Hattiangadi, 2001, describes a holistic review of all of the pays that service members earn when they are away from home and questions whether those pays were adequate or whether a new deployment pay was needed. Among the pays the authors considered were IDP, HFP, and CTZE, as well as the Family Separation Allowance and sea pay. In the review, Koopman and Hattiangadi found several inconsistencies and gaps with respect to IDP and CZTE, some of which remain even today. These gaps suggested to the authors that pays were sometimes applied inconsistently and were not always appropriately structured. For example, they found that, prior to the modern Global War on Terrorism, there were cases of an uneven alignment between IDP zones and combat zones for the purpose of defining CZTE and rules about the accompaniment of dependents. In particular, Oman had been designated an IDP zone, yet service members could bring dependents. In contrast, other locations, such as Qatar, were designated as an IDP zone, but dependents were not allowed—a clearly inconsistent application of a standard, according to Koopman and Hattiangadi. Many of our interviewees said that accompaniment of dependents was a major justification for removing Kuwait from the IDP list in 2014. Koopman and Hattiangadi concluded that service members received unequal benefits, both pecuniary and nonpecuniary, when stationed away from home.

Government Accountability Office Report on Strengthening Combat Pay and Benefits (2006)

In a 2006 report (GAO, 2006), GAO identified three major ways in which the administration of IDP could be improved:

1. maintaining better oversight of existing designations
2. providing clearer guidance on responsibilities and criteria used in the review process
3. minimizing travel that spans multiple months.

Regarding the first point, the report states that, as of 2006, DoD guidance required annual reviews and left the responsibility to the COCOMs to perform them. But the GAO report also states that, in practice, OUSD(P&R) had the responsibility to initiate the reviews and had not done so annually. In the April 2010 version of DoDI 1340.09, DoD set responsibility for initiating reviews with the Principal Deputy Under Secretary of Defense for Personnel and Readiness. In the 2018 version, responsibility lies with the ASD(M&RA) (DoD, 2018). Although the DoD guidance requires periodic review, it does not stipulate the frequency of reviews. As we discussed in the summary of our interviews in Chapter Three, concerns about IDP designations falling out of date are still as present now as they were in 2006.

As for providing clearer policy guidance (the second area for improvement), the 2006 GAO report notes that the IDP review process on paper did not align with the way it was conducted in practice, specifically with respect to the allocation of responsibility. The 2018 DoDI addresses this concern. It defines clear roles and responsibilities for different offices involved in the review process, and our interviewees agreed that the process now works largely as the DoDI states.

The 2006 report also states that there was a lack of a definition of *imminent danger* in both statute and DoD guidance. The report notes that, although DoD used a questionnaire about the factors related to imminent danger, similar to the questionnaire in Appendix 3A of DoDI 1340.09, there was no standard threshold for what constituted an imminent threat. GAO accessed 54 completed questionnaires from between 1998 and 2005 and found evidence that denials are not unusual: In that period, DoD denied requests for Bulgaria, Cyprus, Hungary, Kazakhstan, Romania, and Turkmenistan, and perhaps others. Despite the regularity of both IDP approvals and denials, no consistent threshold of imminent danger was applied to various categories of threats. In particular, GAO recommended establishing a minimum terrorist threat level for IDP eligibility, based on the Defense Intelligence Agency terrorism threat levels of high, significant, moderate, and low. GAO noted that, at the time of the report, adopting a threshold of significant would terminate IDP in 19 areas. It appears that DoD may have adopted this recommendation and set a threshold of a high terrorist threat level, because COCOMs are now required to justify any IDP request for an area with a threat level below high (DoD, 2018, Appendix 3A, § C.3). Our interviewees suggested, however, that *imminent danger* remains a nebulous concept that still lacks a clear set of criteria.

Finally, the 2006 GAO report recommends that DoD "establish departmentwide policies and internal controls that include periodic audits to monitor cross-month travel to ensure that the travel needs to cross calendar months" (GAO, 2006, p. 29). This was in response to the finding that there were a substantial number of trips meriting IDP, CZTE, or both that spanned multiple months, despite being as short as two days long. These trips were not long-term deployments but rather short-term trainings, special missions, or site visits. Similarly, several ships made short-term port visits that spanned two different months. At the time, IDP was not prorated, so each of these trips earned two months' worth of IDP or CZTE. GAO calculated that this issue applied to several thousand individual service members between FY 2003 and FY 2005. Several of our SMEs also noted that these types of payouts would occur, and some interviewees even noted what they viewed as abuses of the system in the form of service members deliberately planning to enter an IDP zone at the very end of one month and very beginning of the next. The excess costs and perceived unfairness of this eligibility standard was eliminated when IDP began to be prorated.

Tenth Quadrennial Review of Military Compensation (2008)

The Tenth QRMC (OUSD(P&R), 2008) proposes a consolidation of special and incentive pays, including the consolidation of HFP, IDP, and HDIP under the umbrella of *hazardous duty pay* (OUSD(P&R), 2008). As already noted, the authorization for this consolidation was enacted by Congress as part of the FY 2008 NDAA, and DoD consolidated the pays when it created hazard pay in January 2018.

In its discussion of the proposed pay consolidations, the Tenth QRMC provides the following rationale for HFP, IDP, and HDIP:

> Hazardous Duty Pay is paid to members serving in dangerous conditions. It is targeted at unpredictable aspects of service, such as deployment to combat zones. This pay is a form of insurance that members know they will receive if their duty situation meets the conditions for eligibility. (OUSD(P&R), 2008, p. 47)

Later in the report, hazardous duty pay is contrasted with assignment duty pay: "While Assignment/Duty Pay would compensate members in anticipation of a burdensome assignment, Hardship/Hazardous Duty Pay would only be paid after the onerous circumstance occurred" (p. 102). The concept of pay as insurance paid after the occurrence of a particular event or circumstance is useful for understanding HFP and IDP as distinct from many other types of special and incentive pays that are meant to ensure that the services meet readiness needs.

The Tenth QRMC also identifies some gray areas regarding HFP and IDP, including many that were raised by our interviewees. Although the QRMC does not provide recommendations on how DoD should resolve such questions, it does identify various alternatives. Three areas addressed in particular are as follows:

1. Variation in pay rate due to level of danger: Should the pay amount vary with the level of danger? The QRMC implies that this would be warranted, although levels of danger are difficult to determine, and a flatter pay scheme may be justified out of feasibility considerations.
2. Variation in pay rate by experience or pay grade: The QRMC acknowledges that the level of danger in a given situation would not vary by tenure or grade. Yet, compensation for danger is computed as a percentage of base compensation, so it will be *perceived* as more or less substantial by individuals with different base salaries and therefore different tenures or grades.
3. Variation in pay rate by service: The QRMC notes that, to capture variation in the amount of risk or danger faced by different service members based on duty assignments and missions, it might be necessary to provide the services with some leeway in setting different rates.

Eleventh Quadrennial Review of Military Compensation (2012)

As part of the Eleventh QRMC (OUSD(P&R), 2012), researchers from the Institute for Defense Analyses wrote a report on combat compensation (Pleeter et al., 2012). They analyzed data from a 2010 survey of service members' perceptions of risks. The study found that, among military personnel who had been in a combat zone, about 30 percent felt that it was much more dangerous than their U.S.-based location, while 22 percent felt that it was no more dangerous. The authors highlight the conclusion that, "not only are risks, as reflected by casualties, quite low in some parts of designated combat zones, but . . . service members know it" (Pleeter et al., 2012, p. 376). Although not every part of a combat zone receives IDP, the conclusion is related to an opinion heard in several of our interviews that IDP does not reflect consistent levels of risk across all designated zones. The authors of the Eleventh QRMC main report recommended that DoD establish one or more pay levels for IDP that are correlated with different levels of threat. Doing so would better align IDP with the varying levels of dangers faced by service members (OUSD(P&R), 2012). The authors also recommended that IDP be separated from HFP—specifically, that the amount paid for IDP should be less than for HFP, given that HFP is paid because of verified exposure to combat.

Government Accountability Office Report on Imminent Danger Pay (2014)

A second GAO report in 2014 focuses on the U.S. Central Command (CENTCOM) area of responsibility, but its conclusions and recommendations apply to IDP more generally (GAO, 2014). After examining the costs of several special pays between 2010 and 2013, GAO found that DoD had not conducted a complete review of IDP in the CENTCOM area since 2007. Yet, over the same period, GAO determined that DoD spent more than $1 billion on IDP in that area alone. Thus, GAO expressed concern about lack of oversight of IDP in this area.

As it turned out, at the time of the 2014 GAO report, CENTCOM had completed a 2011-requested review of IDP in its area of responsibility. It took CENTCOM 18 months to file the necessary documentation, but, in the report, it recommended that several designations be terminated. Unfortunately, the final determination had not been made at the time DoD saw GAO's draft report, so CENTCOM's review was not incorporated into the GAO report or DoD's response. Within just ten days of DoD's response, a final determination was made and several IDP designations in CENTCOM and other areas were slated for termination as of June 2014, with an estimated savings of $150 million per year; see Table B.1 for the areas that were terminated (GAO, 2014, pp. 3–4).[1]

The 2014 GAO report reiterates more pointedly some of the same concerns raised in the 2006 report—namely, the lack of specific guidance on the frequency of periodic reviews and the turnaround time for IDP packets. The 2014 report notes that the most recent modification of DoDI 1340.09 at the time, in April 2010, made the guidance on periodic reviews even less specific than it already was, eliminating the requirement of biennial reviews and instead leaving the frequency unspecified (GAO, 2014, p. 7). Furthermore, the DoDI provided no guidance on how quickly the review process should conclude. GAO's recommendation in 2014 was simple: Revise the DoDI to include specific guidance on the timing of periodic reviews. The 2018 version of DoDI 1340.09 still does not contain such guidance (DoD, 2018).

---

[1] As with the 2006 report, this GAO report provides helpful insight into the content of recommendation packets: CENTOM had recommended a delay in terminating IDP so that troops who were currently assigned to those areas would not experience a sudden decrease in their paychecks. See GAO, 2014, p. 6.

CHAPTER FIVE

Risk-Based Pay in Other Agencies and Militaries

To provide some perspective on the methodology for awarding IDP and HFP and identify alternative practices for doing so, we also considered how a selected set of other organizations award risk-based pay. We considered danger pay for federal civilians, including civilians accompanying military forces, and civilian employees of the United Nations. We then considered risk-related pay for military personnel in Canada and the United Kingdom. The chapter ends with a comparison of the methodology for awarding risk-based pay in these organizations versus in the U.S. military.

U.S. Government Civilian Employees: Danger Pay

Under U.S. Code, Title 5, Section 5928, the U.S. Department of State can establish danger pay for all federal employees stationed in a foreign area. This means that the State Department is responsible for determining danger zones for all civilian government employees and contractors, even those who work outside of the State Department. This includes DoD civilians accompanying U.S. military forces, who may earn danger pay under either Department of State Standardized Regulation (DSSR) Section 652f or Section 652g. Section 652f authorizes danger pay for all civilians, while Section 652g applies only to those accompanying U.S. military forces.

Under Section 652f, danger pay is awarded on the basis of "insurrection, civil war, terrorism, or wartime conditions which threaten physical harm or imminent danger" (U.S. Department of State, 2015). This is the same language used in one of the statutes authorizing IDP (37 U.S.C. § 310). Unlike IDP, danger pay earned under Section 652f is not a flat rate but rather a percentage of regular pay: 15 percent, 25 percent, or 35 percent. Eligibility requires at least four hours of work in one day, including temporary duty assignments.

Under DSSR Section 652g, civilians accompanying U.S. military service members in an IDP-designated area may alternatively earn danger pay at the same flat rate as IDP: $7.50 per day, up to $225 per month. Danger pay under this section may not be paid to anyone who is already receiving danger pay under Section 652f. Therefore, Section 652g may be thought of as extending IDP benefits to civilians in areas that DoD has designated as posing an imminent danger but that the State Department has not.[1]

---

[1] For example, as of March 2019, the State Department has not set a danger pay rate for Niger, but all of Niger is an IDP-designated area. Civilians accompanying military forces in Niger would therefore earn danger pay as IDP, but other civilians would not be eligible for danger pay.

Danger pay is related to two other overseas differentials: post hardship differential and service-needs differential (also known as the difficult-to-staff differential) (see DSSR Section 652d). The post hardship differential is analogous to HDP-L in that it accounts for conditions that are of notably lower standards than would be expected in the United States and that affect the majority of government employees assigned to that location. Unlike HDP-L, the hardship differential is a percentage of the person's basic pay, not exceeding 35 percent (5 U.S.C. §5925(a)). Eligibility requires that individuals also be eligible for living quarters allowances and have spent at least 42 consecutive days at the post (Office of Allowances, undated). The post hardship differential is calculated based on information about the post and includes a certain proportion attributable to political violence and terrorism (the "political violence credit," per Section 652g). Whenever a location is eligible for danger pay, the hardship differential is reduced by the amount attributable to the political violence credit. This policy is similar to the way HDP-L is capped when a location is IDP-designated.

The service-needs differential is paid to service members at posts with especially adverse conditions, where additional compensation is necessary to fill the position. The differential is payable at a maximum of 15 percent of basic pay, but, when combined with danger pay, the two cannot exceed a total rate of 35 percent (5 U.S.C. § 5928). The service-needs differential is more directly analogous to an assignment-based pay than a hazardous duty pay. Thus, it could be used to compensate individuals both when there are adverse (but not hazardous) conditions and when there are hazardous conditions.

Drug Enforcement Agency and Federal Bureau of Investigation

The State Department authorizes danger pay for Drug Enforcement Agency (DEA) and Federal Bureau of Investigation (FBI) employees separately from that for other government employees. Eligibility criteria are the same, but the designated areas and associated rates may differ. Section 151 of Public Law 101-246 states that the Secretary of State must authorize any request by DEA to provide danger pay to any employees of the agency. The overall post differential for DEA employees is then also revised by removing any political violence credit.[2] Section 11005 of Public Law 107-273 amends Public Law 101-246 to include the FBI, so it also has its own danger pay scale.

Civilians Accompanying Military Forces

Civilians accompanying U.S. military forces are eligible for IDP, in the same amount as paid to the military personnel, if they are not already receiving danger pay defined earlier and if the military personnel are IDP-eligible. In such a case, the civilians also cannot be paid any applicable political violence credit unless an exception is made by the State Department.

Civilian Employees of the United Nations

The UN began employing danger pay in April 2012 to replace what used to be hazard pay and extended hazard pay. The allowance is meant to compensate for imminent danger as manifested in any of three conditions: clear and consistent targeting of UN staff or premises as a result of their association with the UN, high risk of becoming collateral damage in a war or

---

[2]   For both DEA and FBI rates, see Office of Allowances, 2017.

active armed conflict, and medical staff being put at risk in nonprotected environments when deployed to deal with World Health Organization–declared emergencies (International Civil Service Commission, 2019). For situations meeting the first two criteria, danger pay is recommended by the UN Under-Secretary-General for Safety and Security; for the third criterion, recommendations come from the Director-General of the World Health Organization. The Chair of the International Civil Services Commission has final say.

The exact amount allowed for danger pay to UN civilian employees varies depending on where staff members are recruited. Internationally recruited staff earn $1,600 per month; staff who are recruited from the location in question earn 30 percent of the net midpoint of the General Schedule salary scale. Payment is prorated on a daily basis if the staff member is in the area for less than a calendar month. Staff receive the pay whenever they are present in the location, regardless of whether they report to duty at a UN office. For example, staff on maternity leave or sick leave are still eligible if they remain in the designated location. It continues to be paid for up to seven calendar days if the staff member takes work-related leave.

Eligibility determinations for each individual duty station are reviewed on a quarterly basis. The International Civil Service Commission maintains the list of eligible locations (see International Civil Service Commission, undated).

Foreign Militaries

Canada: Hardship and Risk Allowances

The Canadian Military assigns a hardship allowance and risk allowances to recognize both conditions and dangers, codified in the Foreign Services Instructions (Chapter 10) of Canada's National Defence Policies and Standards Compensation and Benefit Instructions.[3] The hardship allowance is roughly equivalent to HDP-L, while the risk allowance is roughly equivalent to IDP. However, unlike HDP-L and IDP, both of the Canadian allowances are based on a combination of location and duties. Each mission is assigned a score from one to six, with rates varying by level and the number of accompanying dependents.

The hardship allowance is meant to compensate for living conditions at a specific post. Rates are reviewed at least twice a year by the Departmental Hardship and Risk Committee. The amount paid is based on the total number of points accrued and the number of dependents accompanying the service member. As of April 1, 2018, the rates varied (in Canadian dollars) from $188 per month (for Level I service members with no dependents) to $1,800 per month (for Level IV service members with four or more dependents).

An additional hardship allowance bonus compensates individuals for repeated deployments and is paid as a percentage of the hardship allowance, from 0 to 290 percent based on total points accumulated. Points are accumulated at the rate of one point per month of service on a relevant operation. This means that service members with more experience will have more points than those with less experience, even when at the same location on the same mission. Thus, although the basic hardship allowance would be the same for everyone on a given mission, the total pay after bonuses may differ.

The risk allowance is meant to compensate for risks associated with a specific post, based on the probability of a hazard occurring and the severity of its impact. The Departmental

---

[3] This subsection draws from Government of Canada, 2018.

Hardship and Risk Committee sets risk allowance levels for each mission and reviews each mission at least twice a year. Missions are assigned risk levels from I to IV, and, in a similar situation as the hardship allowance, the exact amount of risk allowance depends on a combination of risk level and number of accompanying dependents. As of April 1, 2018, the risk allowance ranged (in Canadian dollars) from $188 to $1,200 per month.

Both the hardship and risk allowances have a sunset clause built into the timeline for any changes in rates. Increases in the allowances take effect on the first day of the next month, but decreases take effect with a six-month delay.

Canada also pays an operations foreign service premium for service members who are deployed, to "cover expenses not specifically covered by other allowances and benefits" in recognition of service on operations. Like the hardship allowance bonus, this monthly pay depends on points accumulated during deployments and on the number of accompanying dependents. As of April 1, 2018, the pay ranged (in Canadian dollars) from $839 to $3,067.

Canada also offers tax relief for service members. This tax relief is allowed for all income earned by personnel while serving on a deployed mission abroad, up to and including the maximum pay rate of a lieutenant colonel.[4]

United Kingdom: X-Factor

The British military provides a single special pay that is meant to compensate for the overall disadvantages of military life, as compared with typical civilian life. This comprehensive pay is called *X-Factor*, and its rates are recommended by the Armed Forces' Pay Review Body (2018). X-Factor pay does not vary by occupation, only by grade and component. Rates are reviewed and updated every five years, with the most recent review concluding in 2018. X-Factor rates are based on 13 components, which are weighed according to how they affect service members. Three of the 13 components are considered to be positive aspects of military life: job security, training and personal development, and promotion and early responsibility. The other components are considered to be negative aspects of the military, including frequent moves ("turbulence"), danger, hours of work, and stress.[5] Positive factors serve to decrease X-Factor rates, and negative factors increase them.

X-Factor is paid as a percentage of base pay. As of 2018, the rates were 14.5 percent for active and full-time reserve service (full commitment) service members, 5 percent for part-time reserves, and 0 percent for full-time reserve service (home commitment) and university and medical officer cadets.[6] These rates, however, are tapered at higher pay grades; thus, the highest officer pay grades receive a fraction of the maximum cash value of X-Factor received by officers in lower pay grades.

---

[4]   See Clause 110(1)(f)(v) of the Income Tax Act (Revised Statutes of Canada, 2019). Prior to 2017, only certain risk groups of missions were designated for tax relief; see the amendment in the Budget Implementation Act (Statutes of Canada, 2018), No. 1, Clause 9.

[5]   The ten negative components are turbulence (essentially what the U.S. military refers to as permanent changes of station), spousal/partner employment, danger, separation (from home and family), hours of work, stress/personal relationships, leave, autonomy/management control/flexibility, individual/union/collective rights, and travel to work. See Armed Forces' Pay Review Body, 2018, paras. 6.11 through 6.23 and Table 6.1.

[6]   Rates can be found in Armed Forces' Pay Review Body, 2018, Chapter 6. In the full-time reserve service, full commitment appointments fulfill active-duty roles and are fully deployable. Home commitment appointments serve in a single location, except for training exercises or other prespecified duties. See UK Parliament, 1996, Section 24; UK Ministry of Defence, undated.

Although X-Factor includes *danger* as one of its 13 components, it differs from HFP and IDP in both intention and spirit. Unlike HFP and IDP, X-Factor is meant to compensate service members for risks borne in comparison with civilians, not in comparison with other service members. Furthermore, it does not differentiate those risks based on location: The danger component accounts for being deployed or encountering a hazardous situation rather than being located in a dangerous place. As the Armed Forces' Pay Review Body noted in its 2018 report, "X-Factor is not intended to compensate for the particular circumstances that Service personnel face at any one time but instead is aimed at reflecting the balance of advantage and disadvantage averaged out across a whole career" (Armed Forces' Pay Review Body, 2018, p. 79).

To illustrate how X-Factor compensates for average risks over a service member's career, consider the definition of *danger* used for X-Factor versus that for IDP. For X-Factor, danger constitutes "a threat of real or perceived violence; an environment or area which is deemed physically unsafe or uncomfortable for natural, manmade and/or political reasons; danger of death; short or long-term injury to physical or mental health; and injury to oneself or others" (Armed Forces' Pay Review Body, 2018, p. 81). This definition encompasses some aspects of danger as defined for IDP—particularly risks to physical safety posed by the environment or by political unrest—but also accounts for self-injury and mental health. As noted in a 2014 review of X-Factor components, these latter risks are included in the calculation of danger in order to explicitly account for long-term, deleterious effects of combat and deployment, including posttraumatic stress disorder.

The British Armed Forces also pay an unpleasant work allowance analogous to hardship duty pay. The unpleasant work allowance acknowledges "operating in conditions involving an exceptional degree of discomfort or fatigue, or exposure to noxious substances beyond that compensated by X-Factor," as well as duties that are not in the "normal range of military duties and are considered to be of an objectionable, or harrowing, nature" (Armed Forces' Pay Review Body, 2018, p. 45). As of April 1, 2018, unpleasant work allowance paid in three levels: £ 2.73, £ 6.64, or £ 19.64 per day, with each level pertaining to different types of tasks (UK Ministry of Defence, 2019, paras. 16.0105 and 16.0106).

Comparison of Benefits

Table 5.1 provides a comparison of the main elements of risk-based pays for the U.S. military and for the other organizations we considered in this chapter. The columns on the left show the elements for HFP, IDP, HDP-L, and CZTE for U.S. military personnel, and the remaining columns show the elements for the comparison organizations.

With the exception of the UK military, all of the comparison organizations award danger and risk-related pay based on location. The Canadian military also awards risk-related pay based on location, mission, and operational risk. Thus, there is precedent in the Canadian military for awarding risk-related pay based on severity of risk within a specific location.

The United Kingdom's use of X-factor is not location-specific. As mentioned earlier, X-factor compensates for average risks over a service member's career, as well as other factors that make military service different from civilian employment, such as job security. Arguably, X-factor can be thought of as a compensating differential, similar to the differential in the U.S. military that makes average regular military compensation of U.S. military personnel exceed average compensation of civilians with similar characteristics. As discussed in Chapter One, U.S. military pay exceeds the 70th percentile of earnings of similar civilians (Hosek et al., 2018).

One aspect of risk-related pay among U.S. federal employees is that danger pay is a percentage of earnings, whereas IDP and HFP for military personnel are flat monthly amounts. On the other hand, the CZTE benefit for military personnel is related to earnings, given the progressive nature of the federal tax system that increases the marginal tax rate with income. Thus, military personnel also have an element of danger pay related to income.

Table 5.1
Risk-Based Pays and Benefits for U.S. and International Militaries and Civilians

	HFP	IDP	HDP-L	CZTE	Danger Pay (U.S.)	Danger Pay (UN)	Hardship Allowance (Canada)	Risk Allowance (Canada)	X-Factor (United Kingdom)
Eligible population	U.S. military	U.S. military	U.S. military	U.S. military	U.S. civilians	UN staff	Canadian military	Canadian military	UK military
Policy reference	DoDI 1340.09 (DoD, 2018)	DoDI 1340.09 (DoD, 2018)	DoDI 1340.26 (DoD, 2019)	DoDI 1340.25 (DoD, 2010)	DSSR § 652	International Civil Service Commission, 2019	Canada's Compensation and Benefit Instructions 10.3.05 (Government of Canada, 2019)	Canada's Compensation and Benefit Instructions 10.3.07 (Government of Canada, 2019)	Armed Forces' Pay Review Body, 2018
Who determines benefit levels?	U.S. Congress	U.S. Congress	DoD	Statute	Department of State	UN	Departmental Hardship and Risk Committee	Departmental Hardship and Risk Committee	Armed Forces' Pay Review Body
Eligibility basis	Event	Location	Location	Location	Location	Location	Location	Location	Status
Fixed or variable benefit level?	Fixed	Fixed	Variable	Variable	Variable	Variable	Variable	Variable	Variable
What determines variable benefit level?	—	—	Location	Income and rank	Location and income	Country of residence	Location, experience, and number of dependents	Location and number of dependents	Status and rank

Table 5.1—Continued

	HFP	IDP	HDP-L	CZTE	Danger Pay (U.S.)	Danger Pay (UN)	Hardship Allowance (Canada)	Risk Allowance (Canada)	X-Factor (United Kingdom)
Benefit level	$225 per month	$7.50 per day, up to $225 per month	$50, $100, or $150 per month	• Enlisted: all military income tax-exempt • Officer: all military income tax-exempt, up to basic pay plus IDP of the most senior enlisted	• 15%, 25%, or 35% of regular pay in designated locations • Same rate as IDP if accompanying U.S. military in an IDP-designated area that is not otherwise designated for danger pay	• For international staff: $1,600 per month • For local staff: 30% of the net midpoint of the General Schedule salary scale	• Certain amount per month based on operation hardship level (I–IV scale) and number of dependents; ranges from $188 to $1,800 • Personnel with repeated deployments garner an additional percentage, from 0% to 290%	• Certain amount per month based on operation risk level (I–IV scale) and number of dependents; ranges from $188 to $1,200	Additional percentage of base pay as follows: • Active/full-time reserve (full commitment): 14.5% • Part-time reserve: 5% • Full-time reserve (home commitment) and cadets: 0% • Tapered rates apply to the highest officer pay grades

Table 5.1—Continued

	HFP	IDP	HDP-L	CZTE	Danger Pay (U.S.)	Danger Pay (UN)	Hardship Allowance (Canada)	Risk Allowance (Canada)	X-Factor (United Kingdom)
Eligibility criteria	Any one of the following: • Be exposed to, injured by, or killed by hostile fire • Be on duty in proximity to a hostile fire event and exposed to similar dangers of hostile fire (i.e., in a hostile fire area)	Perform duty in a designated area	Perform duty in a designated area for more than 30 consecutive days	Any one of the following: • Be assigned to a combat zone • Be assigned to a QHDA and eligible for HFP or IDP • Be in direct support of operations in Iraq or Afghanistan and eligible for HFP or IDP	Either of the following: • Be assigned to a post designated for danger pay • Accompany the U.S. military in an area designated for IDP but not danger pay	Be present in a designated location (not necessarily assigned for duty)	Be assigned to a designated post	Be assigned to a designated post	Automatically paid as part of service
Contingencies	Cannot earn at the same time as IDP	Cannot earn at the same time as HFP	Capped at $100 per month if earning IDP	In QHDA or direct support capacity, requires HFP or IDP	Raising danger pay lowers the post hardship differential by the same amount				
Continuation of benefits	Up to 1 year during hospitalization associated with a CZTE-eligible duty location	Up to 1 year during hospitalization associated with a CZTE-eligible duty location		Up to 2 years during hospitalization associated with a CZTE-eligible duty location					

CHAPTER SIX

Conclusions and Recommendations

As we noted in Chapter One, our study approach was guided by the questions posed by the FY 2019 NDAA—specifically, what is the current methodology used for awarding HFP and IDP; is it effective; would an alternative approach based on deployments be more appropriate; and in what ways could the HFP and IDP system be improved to address difficulties in implementation? We assessed effectiveness in terms of the overarching criterion that HFP and IDP meet their stated purposes, as highlighted by the Tenth QRMC and developed by Hosek et al., 2019, to insure members against unpredictable and hazardous events. That is, the pays should recognize situations in which service members are unpredictably or involuntarily in harm's way. As insurance, they should vary with the degree of risk, and, furthermore, the pays should be consistently and efficiently implemented and in a timely and transparent way. This chapter provides our conclusions and recommendations.

The Current Methodology Is Effective but Not Considered Efficient

DoDI 1340.09 (DoD, 2018) provides guidance on the setting of pay amounts, the request process for designating a location for eligibility for these pays, and roles and responsibilities. Almost all of the interviewees indicated that the HFP and IDP process is administratively tractable and provides pay to those who are serving in a threatening environment. Although the amount of time between when a member serves in a designated area and the receipt of the pay could be lengthy (as discussed more later), interviewees were not aware of people falling through the cracks and not eventually receiving the pay they were due. Past studies, including two by GAO, have highlighted ways to improve the implementation, overview, and efficiency of the HFP and IDP process. But none of the studies indicated that service members serving in a threatening environment did not receive their combat pay.

In some cases, modern warfare and a more dynamic threat environment may have led to some personnel, such as special operations forces, missing out on IDP, given that their missions are often classified and they are rapidly deployed. However, several SMEs noted that it would be difficult to implement IDP in a way that covers all classified mission areas, and such gaps are filled by other pay mechanisms. First, as an event-based pay, HFP is available if members are subject to hostile fire. Second, in the case of special operations forces, personnel qualify for other special and incentive pays.

That said, some gaps in IDP coverage might be narrowed by incorporating special operations–specific concerns into the IDP request and review process. While our interviewees indicated that geographic COCOMs coordinated with special operations personnel within their area of responsibility as needed, such coordination was on an ad hoc basic.

The Methodology Could Be Improved by Allowing Imminent Danger Pay Rates to Vary with Severity of Threat but Not by Basing Them on Deployments

Several interviewees said that the IDP designation process tends to define geographic locations rather broadly and keep a designation for a long time to minimize "have nots." The result is that the severity of threat can vary widely within an IDP-designated location. The Eleventh QRMC analyzed metrics of danger and found that the correlation between combat compensation and degree of danger had eroded (OUSD(P&R), 2012). The Eleventh QRMC and several interviewees recommended multiple tiers of IDP rates that would be correlated with the severity of threat. IDP would still recognize personnel exposed to lower levels of threat, but the pay rate would be lower than the rate for those exposed to higher levels of threat. One of the interviewees suggested an alternative approach that could achieve the same aim: Reform IDP so that it is always event-based—basically HFP—and define a location around that event. When the event occurs, the dollar amount under this suggested reform would be higher than the current amount and then taper off over time, unless another event occurs within a pre-specified period.

Interviewees uniformly rejected the idea of basing IDP on deployments or other operationally based metrics. Deployments do not necessarily align with risk, so these alternative approaches would not be fair to service members. Those facing no more risk than when in the United States would receive more pay, while those who are in danger but are not considered deployed would not receive the pay. Furthermore, such approaches were deemed infeasible, and maybe even impossible, to implement. Tracking individual locations and missions was considered extremely burdensome from an administrative standpoint. Furthermore, approvals for IDP would have to occur at a much lower level than current policy dictates, but doing so could ignore political factors (e.g., designating IDP in locations that are considered U.S. allies). Finally, interviewees questioned the need for a deployment-related pay, because HFP compensates for exposure to an actual event and can be originated by commanders in the field. Consequently, these interviewees argued that DoD already has a mechanism built into the HFP and IDP process to recognize exposure to danger that is not geographically based.

Other Organizations Provide Some Lessons

In our review of how other organizations, including foreign militaries, award risk-related pay, we found that nearly all of the organizations awarded pay on the basis of only location. The exception was the Canadian military, which bases pay on mission risk and location. Missions are assigned a risk level from I to IV, and the amount of danger pay varies with risk level. Thus, the Canadian military's experience shows that there is some precedence for a danger-related pay that varies with severity of risk.

We found that the criteria for awarding danger pay in the other organizations we examined are similar to the criteria used for IDP (e.g., insurrection, civil war, and terrorism). However, the UN also considers risk to medical staff who are deployed and put at risk in non-protected areas to deal with World Health Organization–declared emergencies. U.S. military personnel, such as reservists, can also face health risks associated with their military service, but DoD guidance does not explicitly include health risk to service members as a criterion for designating IDP.

Implementation of Hostile Fire Pay and Imminent Danger Pay Raises Some Concerns

Our interviewees indicated that the IDP certification process is typically quite lengthy and thus does not respond to changes in the threat environment in a given location in a timely way.[1] Although IDP can be designated retroactively, the long review cycle could have implications for troop morale. The long process is driven by the lengthy list of offices that have a role or responsibility for IDP designation. IDP is considered a highly visible, public, and political decision. In part, the length of the list of offices is driven by concern about creating "haves" and "have nots," especially for those near the geographic borders of the IDP designation. It is also driven by political concerns about paying IDP in countries that are political allies of the United States, as well as by concerns about the interaction of IDP with other factors, such as CTZE and other risk-related pays or benefits. Because of these driving factors, due diligence requires that the process be thorough and inclusive of the various stakeholders.

A contentious issue is the frequency of periodically reviewing IDP-designated locations to ensure that they continue to merit IDP. DoDI 1340.09 requires periodic review without stipulating how frequently such reviews should occur (DoD, 2018). Worldwide reviews have indeed occurred periodically; several SMEs stated that one was nearing completion as this report was being written. Two GAO reports have called for specified periodic reviews rather than ad hoc reviews (GAO, 2006, 2014), and many of the interviewees raised concerns about how long some countries have been on the IDP list and the "bad optics" this can create, especially if the threat to civilians appears to be low (e.g., where cruise ships can dock in an IDP-designated location). The counterargument is that IDP designation undergoes an extensive review, and CCDRs can request removal of a location (as was recently done for Greece), so the list can vary over time.[2] That said, one interviewee remarked that the COCOMs have little incentive to remove locations because, from the perspective of budgeting, IDP costs are borne by the services.

Many interviewees deemed the $225-per-month rate for HFP and IDP to be too low. We could find no evidence of a connection between this pay and recruiting and retention outcomes. But, insofar as HFP and IDP recognize dangerous duty, the value has eroded over time because the dollar amount has remained unchanged since 2003. Furthermore, HFP and IDP are less than the Family Separation Allowance amount of $250 per month. The logic of paying

---

[1] Although the median time for the certification process between 2008 and 2018 was only six months (OUSD(P&R), 2018), one interviewee mentioned two packages that had been in process for years without being resolved.

[2] Furthermore, regarding optics, the counterargument is that threat criteria are relevant only to service members who are performing duties in dangerous areas that are not necessarily a threat to civilians, such as remote mountain regions, deserts, or forests. Because the location of risk can be dynamic within a country, it is not efficient to designate only one region when the location of risk to members in a country is variable.

more to members for being separated from their families than for being in danger escaped the experts with whom we spoke.

Finally, communication within DoD about the progress of IDP packages could be improved to give visibility to commanders about when soldiers, sailors, airmen, and marines can expect to receive retroactive IDP payments. And those who administer IDP require up-to-date and consistent information about IDP designations. The DoD Financial Management Regulation is not always up to date and can be inconsistent with information provided by the Defense Finance and Accounting Service.

Recommendations

From these conclusions, we make the following recommendations. Some of these recommendations would require amending DoDI 1340.09, while others would require action by Congress.

Create tiered rates of IDP based on severity of threat. Setting IDP to reflect different levels of exposure to danger would address inequities among members who currently receive the same pay but face different exposure. It would also respond to the Eleventh QRMC's finding that the connection between danger and IDP has eroded. One potential concern about such as system is that it would require defining *severity of threat*, although interviewees told us that DoD already makes assessments of severity of threat. Another concern is that multiple tiers within an area of responsibility could exacerbate the "haves and have nots" challenge, which in turn might exacerbate the certification timeliness problem. Tracking individual locations within an area of responsibility may also prove burdensome. Another potential concern is that IDP costs could increase. One way to address cost is to set the lowest level of IDP for the least threat below the current $225 per month. Creating tiered rates of IDP based on severity of threat would likely require congressional action.

Increase the current $225 rate for HFP and IDP. IDP should be increased to restore its real value since 2003 and to exceed the $250 per month Family Separation Allowance, at least for some members. Another approach would be to index IDP and HFP to increases in basic pay. Such policies could increase costs. If increasing HFP and IDP is not feasible from a budgetary perspective, an alternative could be to restructure the pays in a cost-neutral way but still base IDP on severity of threat. For example, as part of a tiered IDP system, the lowest-risk IDP rate could fall below $225, while the higher-risk IDP rates could exceed $250 per month. Increasing the IDP rate above $250 would require congressional action because Congress has authorized the pay only up to $250. Increasing the maximum authorized HFP beyond $450 would similarly require congressional action.

Identify whether it is possible to reduce the length of time for IDP certification. IDP certification requires input from a large number of stakeholders, and although any given stakeholder may have limited time to provide input, the overall process is lengthy. DoD should map out the process and assess how long each stage requires for concurrence to identify whether there are ways to streamline the process—for example, by doing some steps concurrently or even eliminating some steps altogether. Identifying whether it is possible to reduce the length of time for IDP certification would require DoD action.

Institutionalize regular periodic reviews of IDP designations. The last worldwide review was in 2014, and a new one is being completed; thus, arguably, reviews could occur regularly every five years. An alternative approach is to include a sunset provision that would decertify a location after five years, unless an HFP event or other information indicates that the location continues to be a threat to service members. Although the periodic reviews do occur and the CCDRs have the flexibility to initiate designation and removal of IDP locations, the commanders' incentives to remove designations may not always align with that of the services. A periodic review would also provide an ongoing basis for justifying the current list of IDP designations. Implementing this recommendation would save money if IDP was no longer paid after the threat at a location has diminished. Institutionalizing a regular periodic review of IDP designations would require DoD action.

Amend DoDI 1340.09 to require the geographic COCOMs to seek input and concurrence from the special operations commander within the COCOM's area of responsibility on packages relevant to special operations forces. This input would occur before the IDP designation request was submitted to the Joint Staff. Doing so would help ensure that information that might currently be missed would be incorporated in the IDP designation request as a matter of course at the point of origination within the geographic COCOMs. Amending DoDI 1340.09 would require DoD action.

Review the criteria in DoDI 1340.09 to assess whether additional risks to service members should be considered as a criterion for designating IDP. For example, biological risks to service members who might be exposed during biological warfare might be added to the list of designation criteria. Insofar as exposure to biological hazards is an aspect of duty rather than a result of hostilities, DoD should consider allowing such members to qualify for hazardous duty pay. Reviewing the criteria in DoD 1340.09 would require DoD action.

Create a capability that would allow IDP administrators across DoD to access up-to-date information. Such a capability could be a website accessible to those involved in the determination process or administration of IDP, or it could be a monthly or quarterly update sent by email to the relevant individuals. This communication would ensure that all involved in the determination process or administration of IDP have a central, authoritative source for IDP information, particularly an up-to-date list of designated countries. Creating this capability would require DoD action.

Wrap-Up

Our main finding is that the methodology for making an HFP or IDP designation or recommendation is relevant and effective. As with most compensation issues, the HFP and IDP process is more complex than it first seems, and changes to it must be done carefully. After completing our study and analysis, we find that support for HFP and IDP remains strong, but some changes to the process are needed to better align the pay to exposure to danger, streamline the process, improve communication, and ensure that the pay continues to serve the needs of service members.

History of Risk Recognition Pay with Zonal Eligibility

In this appendix, we review the history of U.S. military combat and risk-based pay for which eligibility is determined by location. The purpose of this review is to show how the justification for such pay evolved over time, how that justification differs from the typical reasons given for other special pays, and the reasons for current inconsistencies in eligibility and location designation. In addition, the appendix highlights historical studies and criticisms of these pays, which provide context for some the criticisms being voiced today. Unless otherwise noted, the information in this appendix is drawn from Gould and Horowitz, 2012a, except the information in the final section on CZTE, which is drawn from Gould and Horowitz, 2012b. Further pieces of the history of risk-related pay are documented elsewhere, and such sources should be consulted for more detail, including the ways in which these pays were shaped by popular opinion and the mainstream media.

Badge Pay: 1944

Combat pay was first awarded as badge pay during World War II. This pay was provided to combat infantry starting in 1944, providing $10 per month to those with a combat infantry-man's badge (awarded to those in combat service under hostile fire) and $5 per month to those with an expert infantryman's badge (awarded to those with proficiency in training). The justification was to recognize the "hazards and hardships" of frontline service, which were disproportionately borne by infantrymen, as well as to account for the pay discrepancy between infantry and other occupations. In short, the infantry had low morale, and this pay was meant to provide special recognition of their disproportionate share of the war's casualties. In fact, badge pay was not the first attempt to provide such a morale boost; the name derives from actual badges that were originally awarded (starting in 1943) in lieu of pay, but symbolic recognition was not enough. Accelerated promotions and public relations campaigns also failed.

Despite the lack of location-based eligibility, badge pay is the predecessor of today's combat pays in one important sense: It was based on a justification that had never before been used for providing special pay but that continues today in debates surrounding HFP and IDP. Typically, special pays are meant either to incentivize certain types of service (such as taking a particularly risky job or acquiring a scarce skill) or to compensate individuals for particular costs of service (such as injury or family separation). The effectiveness of such pays can be based on careful evaluation: Are the occupations filled? Are enough service members willing to accept duties with certain risks? Recognition, on the other hand, is not tied to measurable outcomes.

Badge pay therefore opened the door to political and military debates regarding the fairness of financial recognition. As Gould and Horowitz, 2012a, p. 211, notes, "Political and military stakeholders must supply the specific policy details. Who is to be recognized? For what risk circumstances? Why is recognition necessary?" Analogous questions may be asked about combat pay as it exists today, and they were asked of badge pay shortly after the war. In fact, the President's Commission on Military Compensation (the Hook Commission) ruled in 1948 that special pays not designed to meet manpower requirements were not warranted in the military pay structure. This resulted in the elimination of badge pay in 1949. The Hook Commission's criticisms would again be relevant when hostilities began on the Korean Peninsula.

Combat Pay: 1952

Combat pay was initiated in 1952 to recognize risks faced by service members in Korea. Like badge pay, it was meant to recognize hazards and hardships faced by particular units. Unlike with badge pay, the original motivation for combat pay was to rectify inequities in hazardous duty pay, not necessarily in morale; for example, submariners and parachutists already received hazard-based compensation, but others facing similar risks did not. This reasoning faced opposition, in large part, because of the reasoning of the Hook Commission: Parachutists needed to be incentivized to volunteer, thereby justifying such pay on manpower grounds. Infantrymen were conscripted, so there was no such reason to provide them with combat pay (Gould and Horowitz, 2012a).

However, the Strauss Commission reaffirmed the morale value of the pay, thereby defending the justification of specialized pay to recognize risk. Eventually, combat pay was justified as a recognition of the frontline hazards and hardships faced by troops in certain units. The pay was created by statute in the Combat Pay Act of 1952.

Also, unlike badge pay—and importantly for subsequent risk-based pay—combat pay was based on conditions faced by particular service members, not on their occupations or training. The pay was awarded to those who served at least six days in designated combat units, as well as those who were wounded or killed by hostile fire (regardless of unit). This was a shift to condition-based eligibility, whereby service members would merit recognition for actual risk experienced as part of their service rather than for their occupation. In the case of combat pay, and unlike modern HFP and IDP, *combat unit* and *risk* were defined by statute: Combat units were frontline ground units,[1] and risk was based on "hazards and hardships"—particularly being subjected to hostile fire.[2] These definitions meant that less than 20 percent of troops deployed to Korea received combat pay (Office of the Assistant Secretary of Defense for Manpower and Reserve Affairs, 1971).

A further precedent set by combat pay was the equalization of benefits across all pay grades. The original plan was for officers to earn double that of enlisted service members, but this faced political opposition based on the notion that all service members' lives were equally

---

[1] This eliminated the Air Force and Navy from eligibility, which was one reason for subsequent debates between the services that led to increasingly expansive eligibility criteria for risk-based pay.

[2] These statutory standards were actually proposed for badge pay by the Tobey-Weiss Senate Bill in 1944, which was never brought to a vote.

valuable. As a result, all eligible service members earned $45 per month. This flat rate across all individuals persists in the statutes authorizing HFP and IDP.

Hostile Fire Pay: 1963

Combat pay expired by statute with the Korean armistice but not before it raised several criticisms from the services. The Navy proposed extending combat pay to all crews of ships exposed to hostile fire, not just to those who were wounded. The Army also noted that troops just to the rear of combat units were also exposed to certain dangers, including guerilla harassment, but were technically ineligible (Commission on Incentive, Hazardous Duty, and Special Pays, 1953). These points raised the possibility of widening the definition of future risk-based pays by using broader definitions of exposure to risk.

Such a shift in eligibility came with the introduction of HFP in the Uniformed Services Pay Act of 1963. HFP was initially modeled after combat pay, restricting eligibility to those serving at least six days in frontline combat units. This was based on the recommendation of the Gorham Commission in 1962, which reiterated the value of recognizing risk and boosting morale, as long as recognition was limited to those on the front lines.

However, the Gorham Commission also recommended that DoD, not Congress, have discretion over the administration of HFP. This meant that DoD could decide who was eligible and why, and this resulted in the 1965 expansion of HFP to all of those serving in Vietnam, with no minimum number of days served. In addition, anyone killed or wounded by hostile fire or other hostile action was granted HFP regardless of whether the service member was in a previously designated HFP zone. This shifted the standard for recognition from condition-based eligibility to zonal eligibility and immediately quintupled the number of HFP recipients; eligibility peaked at 1.25 million in 1968.

Eligibility further expanded in 1968 to those serving on ships and planes outside Vietnam when "one member may be killed or wounded by hostile fire . . . [or] when a hostile act occurs, but no one is wounded or killed" (DoD, 1968). This was in response to attacks on the USS *Liberty* and USS *Pueblo*. DoD also introduced some retroactive designations, including the Korean Demilitarized Zone in 1968, Laos in 1964, and Iran in 1979. These set precedents for further retroactive expansions, such as the expansion of IDP to Mali, Niger, and parts of Cameroon in 2018, which was retroactive to 2017.

The expansion to zonal eligibility was defended by some but criticized by others. The Second QRMC and the House Armed Services Committee argued that zonal eligibility was more practical and equitable because the "front line" was so difficult to define in the new wartime environment of Vietnam. Furthermore, risks were more difficult to estimate than in prior conflicts. In contrast, the President's Commission on the All-Volunteer Force, commonly known as the Gates Commission, recommended reverting to condition-based eligibility. The Commission affirmed the value of recognizing risk but recommended that the pay be tiered based on the frequency of exposure to hostile fire. The Air Force and Navy, among others, were against reverting to condition-based eligibility because their service members had benefited greatly from the expansions to zonal eligibility.[3]

---

[3] The U.S. Marine Corps largely abstained from the debate because it found little justification for any sort of combat pay, yet its service members benefited the second most after the Army's.

Imminent Danger Pay: 1983

HFP never expired, although, besides prisoners of war and those missing in action, very few service members remained eligible after 1974. Because DoD had control over the administration of HFP, it could make new designations for HFP-eligible locations, and the next designation came during the Iran Hostage Crisis in 1979.

The wartime conditions experienced by service members continued to evolve, and so did the argument for expanding HFP eligibility. By 1979, service members were deployed in many more countries than before. This, along with the rise of low-intensity conflicts in the 1980s, prompted further evaluation of the ways in which DoD recognized risk. IDP arose in response to the 1983 bombing of Marine Corps barracks in Beirut and violence against service members in El Salvador. IDP was established in the Department of Defense Authorization Act of 1984 as an expansion of HFP, allowing DoD to grant HFP to "members serving in areas threatening imminent danger," defined "on the basis civil insurrection, civil war, terrorism, or wartime conditions" (Pub. L. 98-94). IDP is therefore a form of HFP for which eligibility is based on risks other than hostile fire.

IDP has been criticized for reasons similar to why HFP was criticized in the Vietnam era. For example, the Fifth QRMC noted that such special pays are not based on a differentiation of various magnitudes or types of risk and are therefore not useful in boosting morale (Office of the Secretary of Defense, 1983).[4] In 2003, an exception to the lack of a tiered pay structure occurred briefly: Service members in Iraq and Afghanistan started earning $225 per month, an increase of the $150 per month that HFP and IDP paid at the time. In 2004, this was made a permanent increase for all HFP and IDP zones, reinstituting a flat pay scale across all eligible zones.

As noted in Chapter Two, the FY 2008 NDAA consolidated HFP, IDP, and HDIP under the umbrella of hazard pay and raised the maximum authorized rate of HFP to $450 per month and that of IDP to $250 per month. The most recent change to HFP and IDP was in the FY 2012 NDAA, which mandated that payments be prorated to $7.50 per day of service in an eligible location, up to the monthly rate of $225.

Combat Zone Tax Exclusion

Combat-related tax exclusions date to World War I, when all service members received a $3,500 tax exclusion regardless of where they served. This exemption was instituted to offset contemporaneous cuts in personal tax deductions that had been made to finance the war. Effectively, Congress raised taxes on civilians to finance the war effort while exempting (most of) those who were serving in the war. Accordingly, the World War I exemption expired when tax rates returned to their prewar levels in 1921.

U.S. leaders had a similar goal with a tax exemption during World War II, which applied to enlisted service members but not to commissioned officers. However, this exclusion was replaced one year later with a $1,500 exclusion for all service members. In 1945, enlisted service members' military income was retroactively excluded from income taxes, and officers were

---

[4]   The Fifth QRMC was released prior to the creation of IDP, so although it does not criticize this pay in particular, the general point applies to IDP.

allowed to exclude a fixed amount. These tax benefits were kept in place until 1949 as a retention incentive in lieu of pay raises.

Contemporary CZTE policy dates to the Revenue Act of 1950, which allowed for enlisted service members to exclude all military income and officers to exclude $200 per month, so long as they were present in a combat zone designated by the President. This policy holds today, with the single major adjustment being the expansion of eligibility based on "direct support" functions. During the Vietnam War, the U.S. Department of Treasury extended CZTE to those in direct support of combat operations, contingent upon receipt of HFP. After the introduction of IDP, it again extended eligibility to those in direct support, contingent upon receipt of IDP. The officers' exemption has also been adjusted: It was first raised to $500 during the Vietnam War and increased again in 1996 to the "maximum enlisted amount," which is interpreted as the pay of the senior enlisted adviser.

Because CZTE is no longer justified by changes in the tax code that are intended to finance wartime operations, it is understood as part of service members' wartime compensation package. This has opened the door for debate over how the CZTE benefit should be implemented—that is, what qualifies as a combat zone. During the Vietnam War, for example, combat operations spread into Cambodia and Laos, and support operations were as far away as Okinawa and Guam, but the combat zone was officially designated as Vietnam (although Laos and Cambodia were added later). Critics argued that serious inequities resulted from this delineation because CZTE depended on the location of a service members' official deployment, while other benefits (such as posthumous exemption for unpaid taxes) were not awarded if the member happened to be killed or injured across the border in Laos.

More recently, combat zones have been defined on a broader geographic basis, avoiding some of these earlier criticisms but raising others. Gould and Horowitz, 2012b, for example, argues that existing combat zones are too broad and do not align with risk according to HFP and IDP designations. For example, the Persian Gulf combat zone included Iraq and Kuwait—areas with actual combat conditions—as well as Saudi Arabia, where combat was expected to spill over; but it also included Qatar, Bahrain, Oman, the United Arab Emirates, and several seas and waterways where troops never experienced combat risks. In fact, the 1993 termination of HFP and IDP in parts of the Persian Gulf showed that OUSD(P&R) deemed the area to pose minimal risk, but the combat zone was never adjusted to reflect this.[5]

Another inequity created by CZTE is the lack of correlation between the size of the benefit and the degree of risk. Because CZTE benefits are largely based on income, it is difficult to argue that it is a fair recognition of risk. This point was made as early as 1967 when the Treasury Department noted that "the exclusion confers its greatest benefits on senior officers and its smallest benefits on the lowest enlisted grades," among other criticisms (Gould and Horowitz, 2012b, p. 286). For example, Pleeter et al., 2012, finds that CZTE benefits do not correlate with actual casualty rates. CZTE-eligible countries with zero casualty rates have the highest average benefits because of the pay grade structure of the units deployed to those countries. Furthermore, junior enlisted service members have the highest incidence of death and injury, as well as the lowest CZTE benefit.

---

[5] Eventually, in 2001, these areas were again designated for IDP, although several of those new designations were later terminated. The Persian Gulf combat zone remains the same as it was in 1991.

Areas Designated for Hostile Fire Pay or Imminent Danger Pay

Table B.1
Areas with HFP or IDP Designation at Any Point Since September 11, 2001, as of June 2019

Area	COCOM	Notes	Start Date[a]	End Date
Land area				
Afghanistan	CENTCOM	Land and airspace above	1 November 1988	Present
Albania	EUCOM	Land and airspace above	22 May 1997	31 March 2002
Algeria	AFRICOM	Land area	7 March 1995	Present
Angola	AFRICOM	Land area	22 June 1992	31 October 2007
Azerbaijan	EUCOM	Land area	9 June 1995	Present
Bahrain	CENTCOM	Land and airspace above	13 June 1997	1 June 2014
Bosnia and Herzegovina	EUCOM	Land and airspace above	22 June 1992[b]	31 October 2007
Burundi	AFRICOM	Land area	29 November 1996	Present
Cambodia	SOUTHCOM	Land area	15 July 1997	31 October 2001
Cameroon	AFRICOM	North and far north	7 June 2017[c]	Present
Chad	AFRICOM	Land area	11 August 2008	Present
Colombia	SOUTHCOM	Land area	1 June 1985	Present
Cote d'Ivoire	AFRICOM	Land area	27 February 2003	Present
Croatia	EUCOM	Land and airspace above	22 June 1992[b]	31 October 2007
Cuba	SOUTHCOM	Guantanamo Bay detention facilities	26 December 2006[d]	Present
Democratic Republic of the Congo	AFRICOM	Land area	29 November 1996[e]	Present
Djibouti	AFRICOM	Land area	31 July 2002	Present
East Timor	INDOPACOM	Land area and airspace up to 1,500 feet	30 September 1999	31 October 2001
East Timor	INDOPACOM	Land area	1 November 2001	31 May 2014
Egypt	CENTCOM	Land area	29 January 1997	Present
Eritrea	AFRICOM	Land area	31 July 2002	Present

Table B.1—Continued

Area	COCOM	Notes	Start Date[a]	End Date
Ethiopia	AFRICOM	Land area	13 September 1999	Present
Georgia	EUCOM	Land area of Georgia and Abkhazia north of 42° N and west of 43° E	28 January 1997	30 July 2002
Georgia	EUCOM	Land area	31 July 2002	31 October 2007
Greece	EUCOM	14 km radius from the center of Athens (38° 1' N, 23° 44' E)	29 January 1997	26 March 2007
Greece	EUCOM	20 km radius from the center of Athens (38° 1' N, 23° 44' E)	27 March 2007	30 November 2017
Haiti	SOUTHCOM	Land area	23 November 1994	31 May 2014
Indonesia	INDOPACOM	Land area	31 October 2001	31 May 2014
Indonesia	INDOPACOM	Land area of the city of Jakarta; the provinces of Central Java, East Kalimantan, and Central Sulawesi; and the Papua region of Aceh province	1 June 2014	Present
Iran	CENTCOM	Land area	4 November 1979	Present
Iraq	CENTCOM	Land area and airspace above (including Basra Oil Terminal)	17 September 1990	Present
Israel	EUCOM	Land area	31 January 2002	Present
Jordan	CENTCOM	Land area	29 January 1997	Present
Kenya	AFRICOM	Land area	31 July 2002	Present
Kosovo	EUCOM	Land and airspace above	22 June 1992[b]	Present
Kuwait	CENTCOM	Land and airspace above	6 August 1990	31 May 2014
Kyrgyzstan	CENTCOM	Land area	19 September 2001	31 May 2014
Lebanon	CENTCOM	Land area	1 October 1983	Present
Liberia	AFRICOM	Land area	6 August 1990	31 May 2014
Libya	AFRICOM	Land and airspace above	19 March 2011	Present
Macedonia	EUCOM	Land and airspace above	22 June 1992[b]	31 October 2007
Malaysia	INDOPACOM	Land area	31 October 2001	31 May 2014
Malaysia	INDOPACOM	Land area of Sabah State	1 June 2014	Present
Mali	AFRICOM	Land area	5 February 2013	30 September 2013
Mali	AFRICOM	Land area	7 June 2017[c]	Present
Montenegro	EUCOM	Land and airspace above	22 January 1992[b]	31 May 2014
Niger	AFRICOM	Land area	7 June 2017[c]	Present

Table B.1—Continued

Area	COCOM	Notes	Start Date[a]	End Date
Oman	CENTCOM	Land area	19 September 2001	31 May 2014
Pakistan	CENTCOM	Land area	29 November 1996	Present
Peru	SOUTHCOM	Land area	1 April 1987	31 December 2001
Philippines	INDOPACOM	Land area	31 October 2001	31 December 2015
Philippines	INDOPACOM	Land area of the island of Mindanao and Sulu Archipelago	5 October 2017[f]	Present
Qatar	CENTCOM	Land and airspace above	7 August 1997	31 May 2014
Rwanda	AFRICOM	Land area	6 October 1997	31 May 2014
Saudi Arabia	CENTCOM	Land and airspace above	2 August 1990	31 May 2014
Serbia	EUCOM	Land and airspace above (including the province of Vojvodina)	22 June 1992[b]	31 May 2014
Sierra Leone	AFRICOM	Land area	18 July 1997	31 October 2007
Somalia	AFRICOM	Land and airspace above	28 September 1992	Present
South Sudan	AFRICOM	Land and airspace above	9 July 2011[g]	Present
Sudan	AFRICOM	Land and airspace above	4 October 1993	Present
Syria	CENTCOM	Land area	31 July 2003	Present
Syria	CENTCOM	Airspace above	21 September 2014	Present
Tajikistan	CENTCOM	Land area	31 March 1997	31 May 2014
Tunisia	AFRICOM	Land and airspace above	9 March 2011	Present
Turkey	EUCOM	Land area encompassing 40-mile radius from the center of Izmir	1 March 1998	24 October 2014
Turkey	EUCOM	Land area, excluding the Turkish Straits (that is, the Dardanelles and the Bosporus) and the Sea of Marmara	29 January 1997	Present
Turkey	EUCOM	Airspace south of 37° 45′ N and east of 43° 00′ E)	1 March 1998	Present
Turkey	EUCOM	Land area east of 33° 51′ E	19 September 2016	Present
Uganda	AFRICOM	Land area	19 January 2000	Present
United Arab Emirates	CENTCOM	Land area	19 September 2001	31 May 2014
Uzbekistan	CENTCOM	Land area	19 September 2001	31 May 2014
Yemen	CENTCOM	Land area	25 May 1999	Present

Table B.1—Continued

Area	COCOM	Notes	Start Date[a]	End Date
Water area				
Arabian Gulf		Including the surface area of the Red Sea, the Gulf of Aden, the Gulf of Oman, and part of the Arabian Sea north of 10° N and west of 68° E	19 September 2001	31 May 2014
Mediterranean Sea		Sea area east of 30° E; only if in connection with Operation Iraqi Freedom	19 March 2003	11 April 2003
Mediterranean Sea		Sea area east of 30° E	12 April 2003	31 July 2003
Mediterranean Sea		The water area of the Mediterranean Sea extending from the North African Coast northward into the Mediterranean Sea and bounded on the east at 26° E longitude, extending north to 34° 35′ N latitude, extending west to the east coast of Tunisia	19 March 2011	Present
Persian Gulf		Water area	1 March 1998	31 May 2014
Somali Basin		Area bounded by the following coordinates: 1110N3-05115E2, 0600N6-04830E5, 0500N5-05030E8, 1130N5-05334E5 and 0500N5-05030E8, 0100N1-04700E1, 0300S3-04300E7, 0100S1-04100E5, 0600N6-04830E5	26 December 2006	Present

SOURCES: Office of the Under Secretary of Defense (Comptroller), 2019, Chapter 10, Summary of Major Changes section; see also the same section in earlier editions of that volume, dating back to 2001, available at Office of the Under Secretary of Defense (Comptroller), undated. For details on what is included in land area, airspace, and water designations, see the notes to Figure 10-1 in those sources.

NOTE: AFRICOM = U.S. Africa Command; EUCOM = U.S. European Command; INDOPACOM = U.S. Indo-Pacific Command; SOUTHCOM = U.S. Southern Command.

[a] For details on memos authorizing each designation, including any relevant renewal of designations, see Office Of the Under Secretary of Defense (Comptroller), 2019, Figure 10-1.

[b] This country was designated when the area was part of Yugoslavia and continued when the country gained independence.

[c] The designation memo was signed March 5, 2018, paid retroactively to June 7, 2017·

[d] The designation memo was signed February 15, 2007, paid retroactively to December 26, 2006.

[e] This country was named Zaire prior to 1997.

[f] The designation memo was signed October 1, 2018, paid retroactively to October 5, 2017.

[g] The designation became effective on the date South Sudan became an independent country, based on the prior designation of Sudan.

Areas with Combat Zone Tax Exclusion Eligibility

Table C.1
Areas with CZTE Eligibility at Any Point Since September 11, 2001, as of June 2019

Area	Details	Start Date	End Date
Combat zone			
Arabian Peninsula (Executive Order 12744)	Bahrain, Iraq, Kuwait, Oman, Qatar, Saudi Arabia, United Arab Emirates, the Persian Gulf, the Red Sea, the Gulf of Oman, the Gulf of Aden, and parts of the Arabian Sea	17 January 1991	Present
Former Yugoslavia (Executive Order 13119)[a]	Albania, Kosovo, Montenegro, Serbia, the Adriatic Sea, and the Ionian Sea north of the 39th Parallel	24 March 1999	Present
Afghanistan (Executive Order 13239)	Afghani land area and airspace, as well as deployments in conjunction with Operation Enduring Freedom	19 September 2001	Present
QHDA			
Balkans (Pub. L. 104-117)	Bosnia and Herzegovina, Croatia, and Macedonia	21 November 1995	Present[b]
Former Yugoslavia (Pub. L. 106-21)[c]	Albania, Kosovo, Montenegro, Serbia, the Adriatic Sea, and the Ionian Sea north of the 39th Parallel	19 April 1999	Present
Sinai Peninsula (Pub. L. 115-97)	Sinai Peninsula	9 June 2015[d]	Present
Direct support eligibility			
Djibouti	Direct support of the Afghanistan combat zone	1 July 2002	Present
Jordan	Direct support of the Afghanistan combat zone	19 September 2001	Present
Jordan	Direct support of the Arabian Peninsula combat zone	19 March 2003	Present
Kyrgyzstan	Direct support of the Afghanistan combat zone	19 September 2001	31 May 2014
Lebanon	Direct support of the Arabian Peninsula combat zone	12 February 2015	Present[e]
Mediterranean Sea	Direct support of the Arabian Peninsula combat zone; limited to the water area east of 30° E	19 March 2003	31 July 2003

Table C.1—Continued

Area	Details	Start Date	End Date
Pakistan	Direct support of the Afghanistan combat zone	19 September 2001	Present
Philippines	Deployments in conjunction with Operation Enduring Freedom	9 January 2002	30 September 2015
Somalia	Direct support of the Afghanistan combat zone	1 January 2004	Present
Somalia	Direct support of the Afghanistan combat zone in the portion of the Somali Basin delimited by the following coordinates: 1110N3-05115E2, 0600N6-04830E5, 0500N5-05030E8, 1130N5-05334E5 and 0500N5-05030E8, 0100N1-04700E1, 0300S3-04300E7, 0100S1-04100E5, 0600N6-04830E5	1 January 2007	Present
Somalia	Direct support of the Afghanistan combat zone in the Somali airspace	1 January 2007	Present
Syria	Direct support of the Afghanistan combat zone	1 January 2004	Present
Tajikistan	Direct support of the Afghanistan combat zone	19 September 2001	31 May 2014[f]
Turkey	Direct support of the Arabian Peninsula combat zone	1 January 2003	31 December 2005
Turkey	Deployments to Incirlik Air Base in direct support of the Afghanistan combat zone	21 September 2001	31 December 2005
Uzbekistan	Direct support of the Afghanistan combat zone	19 September 2001	31 May 2014[f]
Yemen	Direct support of the Afghanistan combat zone	10 April 2002	Present

SOURCES: Office of the Under Secretary of Defense (Comptroller), 2019, Chapter 44, Summary of Major Changes section; see also the same section in earlier editions of that volume, dating back to 2001, available at Office of the Under Secretary of Defense (Comptroller), undated.

[a] See White House, 1999.

[b] CZTE in the Balkans expired with the termination of IDP in 2007.

[c] Executive Order 13119 and Public Law 106-21 designated the same areas as both combat zones and QHDAs.

[d] CZTE was paid retroactively after it was created in 2017.

[e] The Lebanon direct support eligibility is due to expire on February 11, 2020.

[f] This is the date that IDP expired. Personnel in direct support may still receive CZTE if they receive HFP.

Hazardous Duty Incentive Pays

Table D.1
HDIP-Designated Duties, Required Training, and Benefit Amounts, as of June 2019

Duty	Eligibility Requirement	Training Requirement	Maximum Authorized Benefit	DoDI 1340.09 Section
Flying duty	Must participate in 4 hours of aerial flight each month[a]		Aircrew member: Up to $250 per month; non-aircrew member: $150 per month	3.4.c
Parachute duty	Must jump at least once during a 3-month period	Must receive designation as a parachutist or be undergoing such training	Static line parachute jumping: $150 per month; military freefall parachutist: $225 per month	3.4.d
Demolition duty	Must engage in demolition using explosives, in disarming or demolishing explosives, in training to do so, or in experimentation with live explosives	Must engage in demolition as a primary duty or be undergoing such training	$150 per month	3.4.e
Experimental stress duty	Must engage in duties as an experimental test subject in one of the following situations: acceleration or deceleration, thermal stress experiments, low-pressure chambers, or high-pressure chambers[b]		$150 per month	3.4.f
Flight deck hazardous duty[c]	For at least 4 days of flight operations in a month, must be present during flight operations at assigned duty station on the flight deck of a ship from which aircraft are launched and recovered		$150 per month	3.4.g
Duty involving exposure to highly toxic pesticides	Must be assigned to entomology, pest control or management, or a preventive medicine function and perform fumigation duties using any of a list of particular fumigants		$150 per month	3.4.h

Table D.1—Continued

Duty	Eligibility Requirement	Training Requirement	Maximum Authorized Benefit	DoDI 1340.09 Section
Laboratory duty utilizing live dangerous viruses or bacteria	Primary duty must involve work to conduct basic research that utilizes live microorganisms that cause diseases with high potential for mortality and for which no immunization or therapeutic procedures are available		$150 per month	3.4.i
Duty involving toxic fuels and propellants	Primary duty must involve servicing aircraft or missiles with highly toxic fuels or propellants, including testing of aircraft or missile systems[d]		$150 per month	3.4.j
Duty involving handling of chemical munitions	Primary duty must involve direct physical handling of toxic chemical munitions or chemical surety material incident to storage, maintenance, testing, surveillance, assembly, demilitarization, disposal, or escort of said substances[d,e]		$150 per month	3.4.k
Maritime visit, board, search, and seizure (VBSS) duty	Must be assigned to a billet designated as requiring frequent VBSS operations and participate in a minimum of 3 boarding missions during the month (excluding training exercises)	Must be properly trained for VBSS operations	$150 per month	3.4.l
Polar region flight operations duty	Must (1) participate in a takeoff from or landing on the ground in Antarctica or the Arctic Ice-Pack or (2) handle cargo in connection with such aircraft in said regions; event must be certified by an appropriate commander		$150 per month	3.4.m
Weapons of mass destruction civil support (WMDCS) team	Must be serving on an approved active-duty tour in excess of 139 days as part of a DoD-designated and -certified WMDCS team	Must be fully qualified for WMDCS team operations	$150 per month	3.4.n
Diving duty	Must be required to maintain proficiency as a diver by frequent and regular dives, by either diving as a primary duty or serving in an assignment that includes diving other than as a primary duty	Must maintain diving qualifications; payment not made during any period of lapsed qualification	$240 per month	3.4.o

SOURCE: DoD, 2018. The rates and details listed here reflect DoD policy as of January 27, 2018.

[a] Officers are not eligible for flying duty HDIP if they are entitled to aviation incentive pay (37 U.S.C. § 334). Enlisted members are not eligible for flying duty HDIP if they are receiving enlisted flying pay (37 U.S.C. § 353).

[b] For low- and high-pressure chambers, HDIP is also payable to instructor-observers or to inside observer-tenders who conduct hyperbaric treatments.

[c] This HDIP is not payable if another type of HDIP is received in the same period.

[d] Eligibility is based on primary duty assignment and does not imply actual exposure to substances in question.

[e] Handling that is incident to the loading, firing, launching, or field storage of chemical munitions during hostilities is explicitly excluded from eligibility.

Combat-Related Pay and Deployment Tabulations

As discussed in Chapter Two, we wanted to investigate whether there are gaps whereby members were deployed to relevant areas but did not receive HFP or IDP. To do so, we sought to use DMDC administrative personnel data to examine the location and amounts of IDP payments relative to the location and duration of service members' assignments. We found that the administrative pay and location data are not sufficiently detailed to perform such an analysis.

Data Description

The data consist of monthly observations of all military service members in active-duty status in the four DoD uniformed service branches (Air Force, Army, Marine Corps, Navy). For each month in active duty, the data record special pays that were paid out in that month (including type of pay, location associated with the pay, and dollar amount); whether the person was eligible for CZTE during the month (and location meriting eligibility); and location of deployment in the CTS, if applicable.

Because the system records person-months, it is not possible to verify a person's location or pay eligibility for durations of shorter than one month. For this reason, we cannot know, down to the dollar, how much of a particular pay a person should have earned because we do not know exactly how many days he or she spent in a given location. In terms of payouts, we cannot trace the pay in a given month to the time at which that pay was earned. The data will indicate, for example, that a service member was paid HFP or IDP in May 2018 for service in Afghanistan. That service may have occurred in May 2018, or it may have occurred earlier and the pay was issued with a delay. It is also possible that the pay was for multiple months of service and represents a lump sum of back pay. An additional issue is that the DMDC Active Duty Pay File combines HFP with IDP, so when the pay is issued outside of an IDP-designated area, we cannot know whether this reflects an error in the pay file or whether it indicates HFP.

In the case of the CZTE benefit, we do not know the amount of the benefit because it is reflected only in the individual's W-2 form at the end of the tax year; instead, we know only whether some portion of the person's pay in a given month was eligible for tax exemption.

As for CTS data, the system provides an important but limited picture of service members' physical locations. The variables denoting CTS location allow us to verify that a service member was indeed physically present in a given location during a particular month; this, in turn, tells us that the service member should have been eligible for particular pays associated with that location. But CTS, by definition, records only deployments in support of the Global War on Terrorism. In practice, this means that the system accounts for only a fraction of the possible areas that a service member might be located—and, in some cases, such as designated direct support areas, some service members may be in support of the Global War on Terrorism and some may not. In Afghanistan, for example, all service members should be deployed on a contingency operation, and CTS should therefore provide a comprehensive picture of how many individuals were present in the country at a given time. For other locations, however, there is no reliable source of such data.

Comparison of Risk-Related Benefits and Deployment

We looked at the location associated with all HFP and IDP payouts, as well as service members' locations in each month. The former allows us to determine how many months are actually being paid out for a given location, while the latter allows us to tabulate how many should have been paid out. With the caveats noted previously, the two numbers should be roughly equal. Unfortunately, prior to October 2016, the location for all HFP and IDP payouts was coded as "unknown," meaning that we cannot match pay locations to assigned locations. In addition, the data do not reliably record service members' physical locations, except in the case of deployment, so we cannot verify the IDP payouts for locations, such as Niger, that do not appear in the CTS location data because service members are assigned on a nondeployment status.

Table E.1 shows tabulations of individual locations for all active-duty service members in any of the four DoD services since October 2016, when HFP and IDP locations were noted in the data. The table highlights a key limitation of the data: Despite many payments being associated with a location, a large number of locations are still listed as unknown. This prevents a detailed comparison of actual payouts to theoretical eligibility.

As an example, consider Afghanistan. Every service member in Afghanistan should be considered deployed, so the first column should be the total number of person-months spent in the country: 188,503. Everyone in Afghanistan should earn IDP, HDP-L, and CZTE, so the numbers in the other columns should also be 188,503. Instead, the IDP payouts appear to be missing about 88,000 person-months of pay, HDP-L payouts are missing 66,000, and CZTE eligibility is missing 6,000. What cannot be determined is whether these person-months are truly unpaid or whether they are instead grouped in the "unknown" category, in which there are approximately 264,000, 2,000, and 502,000 months, respectively. Figure 2.4 in Chapter Two suggests that the latter is the case, but that cannot be verified.

Table E.1
Total Months of Deployment and Location-Based Special Pays and Benefits for FYs 2017–2018

Area	CTS Deployments (person-months)	IDP Payouts (person-months)	HDP-L Payouts (person-months)	CZTE Eligibility (person-months)
Land area				
Afghanistan	188,503	100,323	122,147	182,141
Albania	0	N/A	5	191
Algeria	9	181	133	22
American Samoa	0	N/A	14	N/A
Angola	0	N/A	6	N/A
Antarctica	22	N/A	4	N/A
Argentina	7	N/A	55	N/A
Armenia	0	N/A	140	N/A
Australia	253	N/A	334	N/A
Azerbaijan	1	166	4	4
Bahamas	0	N/A	529	N/A
Bahrain	72,829	8	11,899	13,428
Bangladesh	0	N/A	53	N/A
Barbados	3	N/A	128	N/A
Belgium	3	N/A	N/A	N/A
Belize	51	N/A	10	N/A
Benin	2	N/A	232	N/A
Bolivia	0	N/A	157	N/A
Bosnia and Herzegovina	75	N/A	109	N/A[a]
Botswana	0	N/A	15	N/A
Brazil	4	N/A	9	N/A
British Indian Ocean Territory	0	N/A	3,073	N/A
Bulgaria	40	N/A	261	N/A
Burkina Faso	361	1	78	1
Burundi	0	151	144	N/A
Cambodia	0	N/A	19	N/A
Cameroon	373	10	689	6
Canada	9	N/A	150	N/A
Central African Republic	24	N/A	12	N/A
Chad	37	607	511	20
Chile	1	N/A	N/A	N/A

Table E.1—Continued

Area	CTS Deployments (person-months)	IDP Payouts (person-months)	HDP-L Payouts (person-months)	CZTE Eligibility (person-months)
China	20	N/A	21	N/A
Colombia	55	829	294	8
Costa Rica	0	N/A	128	N/A
Cote D'Ivoire		168	2	N/A
Croatia	11	N/A	162	N/A[a]
Cuba	0	9	1,930	N/A
Cyprus	1	N/A	164	N/A
Democratic Republic of the Congo	30	332	302	N/A
Djibouti	35,108	17,954	2,753	26,763
Dominican Republic	326	N/A	18	N/A
Ecuador	0	N/A	155	N/A
Egypt	3,577	2,181	3,157	146
El Salvador	35	N/A	1,666	N/A
Estonia	145	N/A	125	N/A
Ethiopia	7	216	15	2
Fiji	0	N/A	28	N/A
France	159	N/A	N/A	N/A
Gabon	0	N/A	28	N/A
Gambia	0	N/A	151	N/A
Georgia	6	N/A	48	N/A
Germany	5,148	N/A	N/A	N/A
Ghana	42	N/A	2	N/A
Greece	724	389	3	24
Greenland	9	N/A	1,583	N/A
Guam	8,238	N/A	N/A	N/A
Guatemala	15	N/A	155	N/A
Guinea	2	N/A	146	N/A
Guyana	0	N/A	123	N/A
Haiti	13	N/A	155	N/A
Honduras	555	N/A	70	N/A
Hong Kong	0	N/A	132	N/A
Hungary	3	N/A	112	N/A

Table E.1—Continued

Area	CTS Deployments (person-months)	IDP Payouts (person-months)	HDP-L Payouts (person-months)	CZTE Eligibility (person-months)
Iceland	56	N/A	2	N/A
India	23	N/A	320	N/A
Indonesia	30	191	26	N/A
Iraq	101,914	60,823	11,220	102,152
Israel	640	2,393	484	420
Italy	1,204	1	N/A	N/A
Jamaica	0	N/A	140	N/A
Japan	3,928	N/A	267	N/A
Jordan	28,289	18,646	17,103	30,140
Kazakhstan	8	N/A	11	N/A
Kenya	1,452	1,142	105	4
Kuwait	160,196	204	96,364	29,837
Kosovo	7,128	3,304	6,097	7,048
Kyrgyzstan	2,439	N/A	116	N/A
Laos	0	N/A	153	N/A
Latvia	18	N/A	4	N/A
Lebanon	1,551	604	516	1,015
Liberia	17	N/A	150	N/A
Libya	23	290	118	38
Lithuania	731	N/A	125	N/A
Macedonia	0	N/A	1	N/A[a]
Madagascar	0	N/A	14	N/A
Malawi	3	N/A	147	N/A
Malaysia	23	1025	182	818
Mali	92	78	262	N/A
Malta	0	N/A	148	N/A
Marshall Islands	0	N/A	220	N/A
Martinique	61	N/A	N/A	N/A
Mauritania	19	N/A	104	N/A
Mexico	12	N/A	21	N/A
Micronesia	0	N/A	274	N/A
Moldova	0	N/A	142	N/A

Table E.1—Continued

Area	CTS Deployments (person-months)	IDP Payouts (person-months)	HDP-L Payouts (person-months)	CZTE Eligibility (person-months)
Mongolia	0	N/A	26	N/A
Morocco	6	N/A	140	N/A
Mozambique	0	N/A	9	N/A
Myanmar	0	N/A	8	N/A
Namibia	0	N/A	141	N/A
Nepal	1	N/A	150	N/A
Netherlands	95	N/A	N/A	N/A
New Zealand	136	N/A	N/A	N/A
Nicaragua	0	N/A	1	N/A
Niger	7,180	1,156	1,544	15
Nigeria	0	N/A	73	N/A
North Korea	0	N/A	65	N/A
Northern Mariana Islands	43	N/A	N/A	N/A
Oman	602	N/A	166	194
Pakistan	1,858	1,130	72	1,224
Palau	80	N/A	5	N/A
Panama	64	N/A	148	N/A
Papua New Guinea	0	N/A	4	N/A
Paraguay	0	N/A	141	N/A
Peru	18	N/A	125	N/A
Philippines	135	0	43	N/A
Poland	268	N/A	135	N/A
Portugal	0	N/A	1,465	N/A
Puerto Rico	422	N/A	N/A	N/A
Qatar	69,081	7	54,664	1,353
Republic of the Congo	0	N/A	144	N/A
Romania	433	N/A	6,401	N/A
Russia	28	N/A	39	N/A
Rwanda	0	N/A	157	N/A
Saudi Arabia	6,785	1	661	766
Senegal	7	N/A	22	N/A
Serbia and Montenegro	0	N/A	3	161

Table E.1—Continued

Area	CTS Deployments (person-months)	IDP Payouts (person-months)	HDP-L Payouts (person-months)	CZTE Eligibility (person-months)
Sierra Leone	2	N/A	142	N/A
Singapore	454	N/A	N/A	N/A
Slovakia	0	N/A	136	N/A
Somalia	5,556	2,083	601	4,271
South Africa	0	N/A	587	N/A
South Korea	3,184	N/A	301,589	N/A
Spain	1,103	N/A	1	N/A
Sri Lanka	0	N/A	42	N/A
Sudan	1	329	26	1
Suriname	0	N/A	3	N/A
Swaziland	0	N/A	141	N/A
Switzerland	3	N/A	N/A	N/A
Syria	13,341	14,281	3,888	17,920
Taiwan	0	N/A	20	N/A
Tajikistan	5	N/A	107	N/A
Tanzania	15	N/A	207	N/A
Thailand	338	N/A	386	N/A
Togo	1	N/A	150	N/A
Tokelau	0	1	N/A	1
Trinidad and Tobago	0	N/A	46	N/A
Tunisia	1,506	1,411	41	613
Turkey	20,644	5,323	3,866	3,973
Turkmenistan	0	N/A	151	N/A
Uganda	487	661	22	6
Ukraine	0	N/A	179	N/A
United Arab Emirates	32,441	6	9,039	1,331
United Kingdom	1,385	N/A	N/A	N/A
United States	831	1	0	1
Uruguay	1	N/A	N/A	N/A
Uzbekistan	0	N/A	124	N/A
Venezuela	0	N/A	2	N/A
Vietnam	23	N/A	121	N/A

Table E.1—Continued

Area	CTS Deployments (person-months)	IDP Payouts (person-months)	HDP-L Payouts (person-months)	CZTE Eligibility (person-months)
Virgin Islands	31	N/A	N/A	N/A
Yemen	186	516	13	115
Zambia	22	N/A	3	N/A
Zimbabwe	0	N/A	145	N/A
Water area				
South Atlantic Ocean	1			
North Pacific Ocean	66			
Gulf of Aden	4,314			
Red Sea	4,809			
Gulf of Oman	1			
Persian Gulf	603			
Savu Sead Sea	957			
Adriatic Sea	2			
Eastern Mediterranean Sea	199			
Other				
Unknown/no location given	242,908	264,125	2,010	502,622

SOURCE: Based on CTS and Active Duty Pay File data from DMDC.

NOTES: The table reflects data from October 2016 through June 2018 (the latest available at the time of writing). N/A = not applicable, indicating that the location was not eligible for the specified pay or benefit during this time frame.

[a] Although the location is technically part of a QHDA, it was ineligible for CZTE during this time frame because it was not designated for IDP.

References

Arizona House of Representatives, H. B. 2795, An Act Amending Section 43-1022, Arizona Revised Statutes; Relating to Individual Income Tax Subtractions, 2006.

Armed Forces' Pay Review Body, *Forty-Seventh Report 2018*, London, Cm 9677, July 2018.

Code of Federal Regulations, Title 26, Section 1.112-1, Combat Zone Compensation of Members of the Armed Forces.

Commission on Incentive, Hazardous Duty, and Special Pays, *Differential Pays for the Armed Services of the United States*, Vol. 1, March 1953.

Congressional Research Service, *Military Pay: Controversy over Hostile Fire/Imminent Danger Pay and Family Separation Allowance Rates*, Washington, D.C., RS21632, October 8, 2003.

———, *U.S. Special Operation Forces (SOF): Background and Issues for Congress*, Washington, D.C., RS21048, March 28, 2019. As of August 12, 2019:
https://fas.org/sgp/crs/natsec/RS21048.pdf

DoD—*See* U.S. Department of Defense.

GAO—*See* U.S. Government Accountability Office.

Gould, Brandon R., and Stanley A. Horowitz, "History of Combat Pay," in Office of the Under Secretary of Defense for Personnel and Readiness, *Report of the Eleventh Quadrennial Review of Military Compensation: Supporting Research Papers*, Washington, D.C.: U.S. Department of Defense, June 2012a, pp. 210–266.

———, "History of the Combat Zone Tax Exclusion," in Office of the Under Secretary of Defense for Personnel and Readiness, *Report of the Eleventh Quadrennial Review of Military Compensation: Supporting Research Papers*, Washington, D.C.: U.S. Department of Defense, June 2012b, pp. 267–299.

Government of Canada, National Defence Policies and Standards, Compensation and Benefits Instructions, Chapter 10: Foreign Services Instructions, October 12, 2018. As of August 12, 2019:
https://www.canada.ca/en/department-national-defence/corporate/policies-standards/compensation-benefits-instructions/chapter-10-foreign-service.html

Hennessy-Fiske, Molly, "Army Says Combat Benefits Will Accompany Fort Hood Purple Hearts," *Washington Post*, April 19, 2015.

Hosek, James, Beth J. Asch, Michael G. Mattock, and Troy D. Smith, *Military and Civilian Pay Levels, Trends, and Recruit Quality*, Santa Monica, Calif.: RAND Corporation, RR-2396-OSD, 2018. As of February 1, 2019:
https://www.rand.org/pubs/research_reports/RR2396.html

Hosek, James, Michael G. Mattock, and Beth J. Asch, *A Wage Differential Approach to Managing Special and Incentive Pay*, Santa Monica, Calif.: RAND Corporation, RR-2101-OSD, 2019. As of February 1, 2019:
https://www.rand.org/pubs/research_reports/RR2101.html

Hosek, James, Shanthi Nataraj, Michael G. Mattock, and Beth J. Asch, *The Role of Special and Incentive Pays in Retaining Military Mental Health Care Providers*, Santa Monica, Calif.: RAND Corporation, RR-1425-OSD, 2017. As of February 1, 2019:
https://www.rand.org/pubs/research_reports/RR1425.html

International Civil Service Commission, "Danger Pay," webpage, undated. As of June 21, 2019:
https://icsc.un.org/Home/DataDangerPay

———, *A Guide to the Mobility and Hardship Scheme and Related Arrangements*, New York: United Nations, March 2019. As of June 29, 2019:
https://icsc.un.org/Resources/HRPD/Booklets/MOBILITYENG.pdf?r=01282499

Kapp, Lawrence, *Operations Noble Eagle, Enduring Freedom, and Iraqi Freedom: Questions and Answers About U.S. Military Personnel, Compensation, and Force Structure*, Washington, D.C.: Congressional Research Service, RL31334, February 16, 2005.

Koopman, Martha E., and Anita U. Hattiangadi, *Do the Services Need a Deployment Pay?* Arlington, Va.: Center for Naval Analyses, CNA Report CRM D0004458.A2, December 2001.

Kusiak, Patrick, *Exclusion from Gross Income During War in Combat Zones*, Washington, D.C.: Office of the Deputy Assistant Secretary of Defense for Military Personnel Policy (Compensation), 1996.

Military.com, "State Tax Information," webpage, undated. As of June 29, 2019:
https://www.military.com/money/personal-finance/taxes/state-tax-information.html

Office of Allowances, "Frequently Asked Questions: Post Hardship Differential," webpage, U.S. Department of State, undated. As of August 12, 2019:
https://aoprals.state.gov/content.asp?content_id=175&menu_id=75

———, "Footnote 'u,'" webpage, U.S. Department of State, June 25, 2017. As of August 12, 2019:
https://aoprals.state.gov/content.asp?content_id=211&menu_id=75

Office of the Assistant Secretary of Defense for Manpower and Reserve Affairs, *Report of the 1971 Quadrennial Review of Military Compensation: Hostile Fire Pay*, 2nd ed., Washington, D.C.: U.S. Department of Defense, December 1971.

Office of the Secretary of Defense, *Fifth Quadrennial Review of Military Compensation: Special and Incentive Pays*, Vol. 3, Washington, D.C.: U.S. Department of Defense, November 1983.

Office of the Under Secretary of Defense (Comptroller), "Archived FMR Chapters," webpage, undated. As of August 12, 2019:
https://comptroller.defense.gov/fmr/archives.aspx

———, *Financial Management Regulation (FMR)*, Vol. 7A: *Military Pay Policy—Active Duty and Reserve Pay*, Washington, D.C.: U.S. Department of Defense, DoD 7000.14-R, June 2019. As of August 12, 2019:
https://comptroller.defense.gov/Portals/45/documents/fmr/Volume_07a.pdf

Office of the Under Secretary of Defense for Personnel and Readiness, *Report of the Ninth Quadrennial Review of Military Compensation*, Vols. I–V, Washington, D.C.: U.S. Department of Defense, March 2002. As of January 31, 2019:
https://militarypay.defense.gov/Portals/3/Documents/Reports/9th_QRMC_Report_Volumes_I_-_V.pdf

———, *Report of the Tenth Quadrennial Review of Military Compensation*, Vol. 1: *Cash Compensation*, Washington, D.C.: U.S. Department of Defense, 2008. As of July 25, 2018:
https://www.hsdl.org/?view&did=713320

———, *Report of the Eleventh Quadrennial Review of Military Compensation: Main Report*, Washington, D.C.: U.S. Department of Defense, June 2012. As of January 31, 2019:
https://militarypay.defense.gov/Portals/3/Documents/Reports/11th_QRMC_Main_Report_FINAL.pdf?ver=2016-11-06-160559-590

———, *Report to Congress on Imminent Danger Pay Determination for the Period 2008 Through 2018*, Washington, D.C.: U.S. Department of Defense, November 2018.

Ohio Revised Code, Title 57, Section 5747.01, Income Tax Definitions.

OUSD(P&R)—*See* Office of the Under Secretary of Defense for Personnel and Readiness.

Pleeter, Saul, Alexander O. Gallo, Brandon R. Gould, Maggie X. Li, Shirley H. Liu, Curtis J. Simon, Carl F. Witschonke, and Stanley A. Horowitz, "Risk and Combat Compensation," in Office of the Under Secretary of Defense for Personnel and Readiness, *Report of the Eleventh Quadrennial Review of Military Compensation: Supporting Research Papers*, Washington, D.C.: U.S. Department of Defense, June 2012, pp. 359–409.

Public Law 88-132, Uniformed Services Pay Act of 1963, October 2, 1963.

Public Law 98-94, Department of Defense Authorization Act, 1984, September 24, 1983.

Public Law 101-246, Foreign Relations Authorization Act, Fiscal Years 1990 and 1991, February 16, 1990.

Public Law 107-273, 21st Century Department of Justice Appropriations Authorization Act, November 2, 2002.

Public Law 104-117, An Act to Provide That Members of the Armed Forces Performing Services for the Peacekeeping Efforts in Bosnia and Herzegovina, Croatia, and Macedonia Shall Be Entitled to Tax Benefits in the Same Manner As If Such Services Were Performed in a Combat Zone, and for Other Purposes, March 20, 1996.

Public Law 106-21, An Act to Extend the Tax Benefits Available with Respect to Services Performed in a Combat Zone to Services Performed in the Federal Republic of Yugoslavia (Serbia/Montenegro) and Certain Other Areas, and for Other Purposes, April 19, 1999.

Public Law 108-11, Emergency Wartime Supplemental Appropriations Act, April 16, 2003.

Public Law 108-136, National Defense Authorization Act for Fiscal Year 2004, November 24, 2003.

Public Law 108-375, Ronald W. Reagan National Defense Authorization Act for Fiscal Year 2005, October 28, 2004.

Public Law 110-181, National Defense Authorization Act for Fiscal Year 2008, January 28, 2008.

Public Law 112-81, National Defense Authorization Act for Fiscal Year 2012, December 31, 2011.

Public Law 115-97, An Act to Provide for Reconciliation Pursuant to Titles II and V of the Concurrent Resolution on the Budget for Fiscal Year 2018, December 22, 2017.

Public Law 115-232, John S. McCain National Defense Authorization Act for Fiscal Year 2019, August 13, 2018.

Revised Statutes of Canada, 1985, c. 1, 5th Supp., Income Tax Act, June 17, 2019.

Statutes of Canada, 2018, c. 12, Budget Implementation Act, 2018.

UK Ministry of Defence, "Glossary of Terms and Abbreviations: Defence Statistics (Tri-Service)," undated. As of August 12, 2019:
https://assets.publishing.service.gov.uk/government/uploads/system/uploads/attachment_data/file/569187/Tri-Service_Glossary_-_Nov16.pdf

———, Tri-Service Regulations for Expenses and Allowances, London, JSP 752, July 2019. As of August 12, 2019:
https://assets.publishing.service.gov.uk/government/uploads/system/uploads/attachment_data/file/772297/20190101-JSP_752_v37.pdf

UK Parliament, Reserve Forces Act 1996, May 22, 1996.

U.S. Code, Title 5, Section 5928, Danger Pay Allowance.

U.S. Code, Title 26, Section 112, Certain Combat Zone Compensation of Members of the Armed Forces.

U.S. Code, Title 37, Pay and Allowances of the Uniformed Services.

U.S. Department of Defense, Special Pay for Duty Subject to Hostile Fire, DoD Directive 1340.6, August 1, 1968.

———, Combat Zone Tax Exclusion (CZTE), Washington, D.C., DoD Instruction 1340.25, September 28, 2010. As of August 12, 2019:
https://www.esd.whs.mil/Portals/54/Documents/DD/issuances/dodi/134025p.pdf

———, Assignment and Special Duty Pays, Washington, D.C., DoD Instruction 1340.26, Change 1, January 11, 2019. As of August 12, 2019:
https://www.esd.whs.mil/Portals/54/Documents/DD/issuances/dodi/134026p.pdf?ver=2019-01-11-075015-873

———, *Hazard Pay (HzP) Program*, Washington, D.C., DoD Instruction 1340.09, January, 26, 2018. As of August 12, 2019:
https://www.esd.whs.mil/Portals/54/Documents/DD/issuances/dodi/
134009p.PDF?ver=2018-01-30-123041-040

U.S. Department of State, Standardized Regulation, Chapter 650, Danger Pay Allowance, 2015.

U.S. Government Accountability Office, *Military Personnel: Actions Needed to Strengthen Management of Imminent Danger Pay and Combat Zone Tax Relief Benefits*, Washington, D.C., GAO-06-1011, September 28, 2006.

———, *Imminent Danger Pay: Actions Needed Regarding Pay Designations in the U.S. Central Command Area of Responsibility*, Washington, D.C., GAO-14-230R, 2014.

U.S. House of Representatives, H.R. 1588, National Defense Authorization Act for Fiscal Year 2004, May 22, 2003.

White House, *Designation of Arabian Peninsula Areas, Airspace, and Adjacent Waters as a Combat Zone*, Executive Order 12744, January 21, 1991.

———, *Designation of Federal Republic of Yugoslavia (Serbia/Montenegro), Albania, the Airspace Above, and Adjacent Waters as a Combat Zone*, Executive Order 13119, April 13, 1999.

———, *Designation of Afghanistan and the Airspace Above as a Combat Zone*, Executive Order 13239, December 21, 2001.